MAKING CHEMISTRY RELEVANT

MAKING CHEMISTRY RELEVANT

STRATEGIES FOR INCLUDING ALL STUDENTS IN A LEARNER-SENSITIVE CLASSROOM ENVIRONMENT

Edited by

Sharmistha Basu-Dutt

WILEY

A JOHN WILEY & SONS, INC., PUBLICATION

For general information on our other products and services or for technical support, please contact our Customer Care Department within the United States at (800) 762-2974, outside the United States at (317) 572-3993 or fax (317) 572-4002.

Wiley also publishes its books in a variety of electronic formats. Some content that appears in print may not be available in electronic formats. For more information about Wiley products, visit our web site at www.wiley.com.

Library of Congress Cataloging-in-Publication Data:

Basu-Dutt, Sharmistha.
  Making chemistry relevant : strategies for including all students in a learner-sensitive classroom environment / Sharmistha Basu-Dutt.
      p. cm.
  ISBN 978-0-470-27898-7 (cloth)
  1. Chemistry–Study and teaching.   I. Title.
  QD40.B32 2010
  540.71–dc22

                              2009036210

Printed in the United States of America.

10 9 8 7 6 5 4 3 2 1

■ CONTENTS

Julie K. Bartley, PhD, Associate Professor, Department of Geology, Gustavus Adolphus College, St. Peter, MN

Sharmistha Basu-Dutt, PhD, Associate Professor and Director of Engineering Studies, Department of Chemistry, University of West Georgia, Carrollton, GA

Anne K. Bentley, PhD, Assistant Professor, Department of Chemistry, Lewis & Clark College, Portland, OR

Gautam Bhattacharyya, PhD, Assistant Professor, Department of Chemistry, Clemson University, Clemson, SC

Thomas D. Crute III, PhD, Professor, Department of Chemistry, Augusta State University, Augusta, GA

Laura DeLong Frost, PhD, Associate Professor, Department of Chemistry, Georgia Southern University, Statesboro, GA

Bill Donovan, PhD, Associate Professor and Director of Freshman Chemistry, Department of Chemistry, University of Akron, Akron, OH

Laura Post Eisen, PhD, Coordinator of Women in Science, Health and Medicine cohort of the Elizabeth Somers Women's Leadership Program, and Assistant Professor, Department of Chemistry, George Washington University, Washington, DC

Julie Ellefson, MEd, Associate Professor and Chairperson, Department of Chemistry, Mathematics and Science Division (M/S), William Rainey Harper College, Palatine, IL

Erik M. Epp, MS, Graduate Student, Department of Chemistry, Purdue University, West Lafayette, IN

Victoria J. Geisler, PhD, Associate Professor, Department of Chemistry, University of West Georgia, Carrollton, GA

Farooq A. Khan, PhD, Associate Professor, Department of Chemistry, University of West Georgia, Carrollton, GA

Gail Marshall, EdD, Assistant Professor, Department of Curriculum and Instruction, University of West Georgia, Carrollton, GA

Maria Oliver-Hoyo, PhD, Associate Professor, Department of Chemistry, North Carolina State University, Raleigh, NC

Bonnie Rinard, MS, Senior Research Associate, Center for Occupational Research and Development (CORD), Waco, TX

Cianán B. Russell, PhD, Program Associate, Center for Workshops in the Chemical Sciences (CWCS), School of Chemistry and Biochemistry, Georgia Institute of Technology, Atlanta, GA

Amy M. Shachter, PhD, Associate Professor, Department of Chemistry and Biochemistry, Santa Clara University, Santa Clara, CA

S. Swamy-Mruthinti, PhD, Professor, Department of Biology, University of West Georgia, Carrollton, GA

Gabriela C. Weaver, PhD, Professor and Director of the Center for Authentic Science Practice in Education (ASPiE), Department of Chemistry, Purdue University, West Lafayette, IN

Mark Whitney, PhD, Manager of Publication Services, Center for Occupational Research and Development (CORD), Waco, TX

Donald J. Wink, PhD, Professor, Department of Chemistry , University of Illinois at Chicago, Chicago, IL

Learning is a complex and multifaceted process that can be facilitated in a classroom environment that is challenging, yet flexible, allowing students to find meaningful relationships between abstract ideas and practical applications in the context of the real world. Creating inclusive learning environments that connect to a diverse student population opens doors of opportunity for students throughout their lives. Students' interest and achievement in academics improve dramatically when they make connections between what they are learning and the potential uses of that knowledge in the workplace and/or in the world at large. This book presents a unique collection of strategies that have been used successfully in chemistry classrooms to create a learner-sensitive environment that enhances academic achievement and the social competence of students.

In Chapter 1, Donald Wink reviews some of the rationales for teaching chemistry in a connected manner. Three different approaches are considered, with specific examples from the recent chemistry education literature. The main philosophical point is that knowledge is stronger when it is based on connections between the learners and the material. Rationales for connected teaching in chemistry also arise from understandings of cognition, specifically by considering how they construct meaning within their own thinking. And sociological issues arise when considering how to connect the learner to chemistry because of goals of inclusion and diversity in the enterprise of teaching chemistry.

Chapter 2 introduces the ideals of Science Education for New Civic Engagement and Responsibilities (SENCER), a national dissemination project supported by the National Science Foundation and the National Center for Science and Civic Engagement, focusing on connecting basic chemical principles to complex, public issues through campus- and community-based research projects as well as course content and structure. Amy Shachter describes a SENCER model course that uses students' active participation in civic engagement as a means of learning science.

The Center for Occupational Research and Development (CORD), a national nonprofit organization, provides innovative changes in education to prepare students for greater success in careers and in higher education. In Chapter 3, Mark Whitney and Bonnie Rinard elaborate a laboratory-centered, hands-on activities-based CORD module that emphasizes the importance of teaching science in the context of major life issues surrounding work, home, society, and environment.

Combating terrorism and preserving our environment are two critical issues that are of national and global interest today. Laura Eisen presents an interdisciplinary course on the science of terrorism in Chapter 4. The course is intended to increase the scientific literacy of nonscience majors who will use this knowledge to make sociological, economic, or political decisions related to homeland security in their future careers. In Chapter 5, Gautam Bhattacharya shows how Green Chemistry has taken the center stage in the new millennium as the sustainability of the human race has become the most important scientific, economic, social, and political challenge in the world. Examples presented help students understand that, when used in an environmentally and ecologically conscientious manner, chemicals and chemistry produce many of the staples of human existence.

In Chapter 6, Gail Marshall discusses some general features of a variety of commonly used student-centered, active learning pedagogies in science education. Laura DeLong Frost focuses on reorganizing the content in allied health chemistry, making it more interesting to students and the incorporation of the process-oriented guided inquiry learning (POGIL) approach in Chapter 7.

The development and implementation of the Working with Chemistry (WWC) Laboratory Program is highlighted by Julie Ellefson in Chapter 8. Specific WWC modules that contain sets of experiments linked by a scenario describe a situation/problem that a professional who uses chemistry regularly but is not a chemist may encounter in the context of his or her work.

In Chapter 9, a team of interdisciplinary faculty comprising Julie Bartley, Sharmistha Basu-Dutt, Victoria Geisler, Farooq Khan, and S. Swamy-Mruthinti, describe three freshmen seminars to introduce students to the contextual relevance of introductory science and mathematics courses, providing a "keystone" experience that enables them to engage in science and mathematics within the context of their professional goals. The seminars provide an intellectually exciting experience for entering science, technology, engineering, and mathematics (STEM) students using active learning methods, multidisciplinary experiences, and skill development in the context of real-life problems.

Cianán Russell, Anne Bentley, Donald Wink, and Gabriela Weaver describe selected modules from The Center for Authentic Science

Practice in Education (CASPiE) program in Chapter 10. These modules provide first- and second-year students with access to research experiences as part of the mainstream general and organic chemistry curriculum; create a collaborative "research group" environment for students in the laboratory; provide access to advanced instrumentation for all members of the collaborative to be used for undergraduate research experiences; and create a research experience that is engaging for women and for ethnic minorities and appropriate for use at various types of institutions, including those with diverse populations.

In Chapter 11, Maria Oliver-Hoyo introduces a variety of unique multisensorial experiments that are designed to use other senses in addition to eyesight when studying chemical processes and performing chemical techniques. Besides adding richness to the chemistry experience of all students, these sensorial experiments provide an opportunity to integrate visually impaired students into the laboratory experience in an active and independent manner.

Entertainment and variety are universally sought human experiences, the thirst for which can be exploited in the classroom, with the help of hypermedia. In Chapter 12, William Donovan illustrates the use of interactive websites and response systems to increase their understanding of chemistry concepts alone and in collaboration with other students, tutors, and teaching assistants (TAs), and their instructor. Erik Epp and Gabriela Weaver present the development and use of a *Physical Chemistry in Practice* DVD in Chapter 13 that combines video, audio, and graphical and textual information in an interface. With the help of hypermedia, students can choose the order and pace of accessing information, giving them great flexibility and allowing them to work in the way that they learn best. Successful puzzles and games in the context of learning utility are demonstrated by Thomas Crute in Chapter 14.

This book is a synthesis of work by many passionate faculty members who possess similar goals of creating an inclusive classroom environment for their students. We hope that novice and experienced faculty will find valuable ideas in this book and be able to adapt these or to develop their own strategies to enhance the learning experience of all their students. We welcome your comments and suggestions for improving this book.

SHARMISTHA BASU-DUTT

Carrollton, GA

July 2009

Sharmistha Basu-Dutt is currently Associate Professor of Chemistry and the Director of Engineering Studies at the University of West Georgia. She obtained her BS in Chemical Engineering from Jadavpur University, Kolkata, India and her PhD in Chemical Engineering from Wayne State University, Detroit, MI. She was awarded the 2007 University System of Georgia Regents Excellence in Teaching Award for her innovative teaching techniques, which include collaborative and cooperative learning, inquiry-based learning, integrative teaching, and activity-based interdisciplinary approaches. She leads  several state-wide programs that help college and K-12 students as well as their teachers find meaningful connections between abstract ideas in the sciences and practical applications in the engineering world.

Contact:
Sharmistha Basu-Dutt, PhD
Associate Professor and Director of Engineering Studies
University of West Georgia
1601 Maple Street
Carrollton, GA 30118
Phone: (678) 839-6018
E-mail: sbdutt@westga.edu

Philosophical, Cognitive, and Sociological Roots for Connections in Chemistry Teaching and Learning

DONALD J. WINK

University of Illinois at Chicago, Chicago, IL

INTRODUCTION

This chapter reviews some of the rationales for teaching chemistry in a connected manner. Three different approaches are considered, with specific examples from the recent chemistry education literature. The first rationale comes from philosophy, where knowledge requires connections among experience, inquiry, and the material. A rationale for connected teaching in chemistry also arises from psychology, specifically by considering how students' cognition allows them to construct meaning. And finally, since learning is about the learner's relationship with society, sociological issues arise because of goals of inclusion and diversity in the enterprise of teaching chemistry.

Through all of this, I will try to relate the different points of this discussion to particular contemporary ideas in chemistry and science education. As such, all of the work can be summed up by the following rationale for inquiry teaching, taken from the National Research Council report *Inquiry and the National Science Education Standards* (National Research Council 2005):

> Through meaningful interactions with their environment, with their teachers, and among themselves, students reorganize, redefine, and

From Hamilton, Edith: *The Collected Dialogues, Including the Letters.* © 1961 Princeton University Press, 1989 renewed. Reprinted by permission of Princeton University Press.

replace their initial explanations, attitudes, and abilities. An instructional model incorporates the features of inquiry into a sequence of experiences designed to challenge students' current conceptions and provide time and opportunities for reconstruction, or learning, to occur.

Instructional models ... seek to engage students in important scientific questions, give students opportunities to explore and create their own explanations, provide scientific explanations and help students connect these to their own ideas, and create opportunities for students to extend, apply, and evaluate what they have learned.

THE EPISTEMOLOGY OF CONNECTION

To start our inquiry, we step back and consider where these ideas come from by going back to one of the most fundamental ideas in teaching: Socratic teaching, presented in Plato's *Meno* (1961). On its surface, *Meno* is a dialogue about the teaching and learning of virtue or, rather, the problem that it seems to some that virtue cannot be taught because virtuous people do not always have virtuous children. To discuss this Plato presents a discussion of how learning occurs generally, through the agency of Socrates' questioning a slave boy about a simple mathematical problem: the length of the side of a square with an area of 8. Socrates uses questions and some simple illustrations, eliciting explanations, predictions, and conclusions from the boy. Prior to the beginning of this excerpt, Socrates has guided the boy to indicate that a square that has sides of 2 ft has an area of 4 ft (Fig. 1.1a).

Socrates: And how many feet is twice two? Work it out and tell me.
Boy: Four.
Socrates: Now could one draw another figure double the size of this, but similar, that is, with all its sides equal like this one?
Boy: Yes.
Socrates: Now then, try to tell me how long each of its sides will be. The present figure has a side of two feet. What will be the side of the double-sized one?
Boy: It will be double, Socrates, obviously.
Socrates: You see, Meno, that I am not teaching him anything, only asking. Now he thinks he knows the length of the side of the eight-foot square.

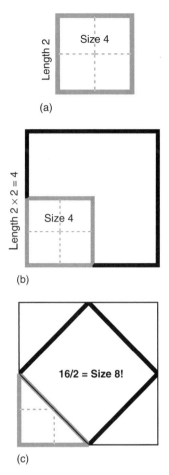

FIGURE 1.1 Diagrams for the geometry inquiry in *Meno*.

Meno: Yes.

Socrates: But does he?

Meno: Certainly not.

Socrates: He thinks it is twice the length of the other.

Meno: Yes.

Socrates: Now watch how he recollects things in order—the proper way to recollect. You say that the side of double length produces the double sized figure? It must be equal on all sides like the first figure, only twice its size, that is, eight feet. Think a moment whether you still expect to get it from doubling the side.

Boy: Yes, I do.

Socrates: Well now, shall we have a line double the length if we add another of the same length at this end?

Boy: Yes.

Socrates: It is on this line then, according to you, that we shall make the eight-foot square, by taking four of the same length?

Boy: Yes.

Socrates: Let us draw in four equal lines using the first as a base. Does this not give us what you call the eight-foot figure? (Fig. 1.1b)

Boy: Certainly.

Socrates: But does it contain these four squares, each equal to the original four-foot one?

Boy: Yes.

Socrates: How big is it then? Won't it be four times as big?

Boy: Of course.

Socrates: And is four times the same as twice?

Boy: Of course not. (Fig. 1.1c)

Socrates: What do you think, Meno? Has he answered with any opinions that were not his own?

Meno: No, they were all his.

Socrates: Yet he did not know, as we agreed a few minutes ago.

Meno: True.

Socrates: But these opinions were somewhere in him, were they not?

Meno: Yes.

Socrates: So a man who does not know has in himself true opinions on a subject without having knowledge.

Meno: It would appear so.

Socrates: At present these opinions, being newly aroused, have a dreamlike quality. But if the same questions are put to him on many occasions and in different ways, you can see that in the end he will have knowledge on the subject as accurate as anybody's.

Meno: Probably.

Socrates: This knowledge will not come from teaching but from questioning. He will recover it for himself.

Meno: Yes.

Socrates: And the spontaneous recovery of knowledge that is in him is recollection, isn't it?

Here and elsewhere in the dialogue, there is no place where we can find Socrates providing direct instruction in the form of "here is the answer." But the boy does come to the correct answer. In the dialogue, Socrates draws the conclusion that the boy must have *already* known the answer, remembered from a previous life since he had not learned it in this life. To avoid a recursion problem, Plato also asserts that the knowledge does get a start, though through divine inspiration, not from the world, at some point in the soul's journeys through the world.

The knowledge theory of *Meno* represents a "myth of remembering," and it is pretty clear that this remembering was what Plato really believed, along with a belief in reincarnation. This is somewhat surprising, then, to find Socratic teaching so popular now, since it is unlikely many believe in reincarnation, including the specific memory of knowledge from previous lives.

However, if we look a bit closer, we can see something else embedded in the dialogue. As the quotation illustrates, Socrates does not "tell," but he does "guide" the boy to look at and to do certain things, including constructing diagrams of a particular type. The boy is not a blank slate from which Socrates draws remembered knowledge. It is true that the knowledge does come from inside the boy's head. But it is knowledge that originates in an attempt to explain the features of the world. This knowledge is then tested by making predictions about the world, observing results, and, when the prediction fails, revising the explanation. While Plato may tell us that this is a result of remembering, it is easy for us to see this process as one in which *experience*, not memory, is the source of knowledge.

Thus, this interpretation of *Meno* (and analysis of Socratic teaching) is that we do not come to know about the natural world through divine inspiration or by conjuring memories of an earlier life. Rather, we come to know because of deliberate and guided experiences with the world, including specific testing of theories and predictions.[1]

For more than two millennia after Plato, the philosophical discussion of epistemology focused on how to get at truth, generally taken to be

---

[1]The reader should know that for Plato, experience could be a teacher, but because we only experience specific things, it could never be trusted. That prejudice is, perhaps, why Plato formulated the myth of remembering and of divine inspiration, since the divine was eternal and universal, in contrast to the unreliable experiences of our world. Aristotle, in contrast, did introduce and use an epistemology that gave priority to perception and experience as necessary precursors to the understanding of abstract principles (Scott 2007). Aristotle did not give us a script for teaching in the way Plato did with the *Meno*.

a single thing that was, in theory, attainable by inquiry. Whether this derived from empirical or idealist perspectives did matter, but the idea of a single outcome for inquiry, properly done, was unquestioned, at least in the Western traditions. Even the seeming overthrow of theistic perspectives in the Enlightenment only gave rise to ideas that sought to locate the discovery of truth as outside of revelation. It was not until the growth of scientific materialism, in particular spurred by Darwin's theory of descent by natural selection, that questions of removing a fixed truth from epistemology seemed possible, including the work of the American pragmatist movement. This was summarized in the lectures that William James gave on the subject. In these lectures, he makes clear that truth is rooted in how the knower sees a given proposition or method as *useful* to achieve an object of value.

As he put it in his lecture "Pragmatism's Conception of Truth," "True ideas are those that we can assimilate, validate, corroborate, and verify. False ideas are those that we cannot." And, shortly later, "The truth of an idea is not a stagnant property inherent in it. Truth happens to an idea. It becomes true, is made true by events." This, though, invites a further question: What events can make something true? To answer this, James reflects at length on two realities that constrain thought: the realities of matters of fact and the realities of ideal relations as they can be formulated by the mind; events in our experience of the world and events in our thinking contribute to making things true. But he introduces a third kind of event, one critical to pragmatism: agreement of ideas with what we already understand. Our prior stance, he suggests, shapes both what we experience and what we think. This stance is both intellectual and personal, rooted in who we are as individuals, with particular needs and goals.

An important consequence of this mode of thinking is the recognition that the method of obtaining data, of reasoning, and of setting goals is bound up in what will be true in a given situation. The pragmatists, therefore, opened the door for the recognition that what will be true depends on the methods of truth setting. Rather than see this as a hopeless circular argument, they leave it circular but embed it in a *relation* between knowers and reality, where the outcome is not a fixed truth but is a set of useful (i.e., pragmatic) truths.

The interpretation of truth and knowledge as derived from purposeful reflection on experience and as anchored in utility has been significantly updated by two recent movements. One is the development of the epistemological consequences of the feminist movement to challenge the privileges associated with certain political, sociocultural, or pedagogical structures (Brickhouse 2001). Feminist critiques of science began with considering the reasons for the paucity of women in science, technology, engineering, and mathematics (STEM) fields but in the 1980s, led by thinkers such as Sandra Harding, advanced to other considerations. This included the role of sexism and other forms of bias in deciding what science is done, and the impact of sexist imagery on the methods of science in treating nature as an object to be mastered. However, the project also led feminist philosophers to raise questions of what it means to "know" in broad philosophical terms. Given the general feminist perspective that our actions, including acts of knowledge, are linked to our individuality, this epistemological perspective joined others that questioned the privilege of science as a method and as a body of knowledge. Harding and others sought a way to formulate a reliable perspective that replaced absolute systems. This led to the concept of feminist standpoint theories, summarized by Donna Haraway (1988):

> We do not seek partiality for its own sake, but for the connections and unexpected openings situated knowledges make possible. The only way to find a large vision is to be somewhere in particular. The science question in feminism is not about objectivity as positioned rationality. Its images are ... the joining of partial views and halting voices into a collective subject position that promises a vision of the means of ongoing finite embodiment, of living within limits and contradictions, *i.e.*, of views from somewhere.

It is easy to take this goal of a "collective subject position" and to equate it with a single privileged position of knowing. As Harding (2004) argues, however, a key component of standpoint theory is that knowledge always comes from a set of "somewheres" that are constantly in flux and, in fact, are in relation with one another. This means that the goal of a collective position is not to find one place for knowledge but to find the one place that persons have assembled at a given time and place, understanding that this may or may not be the same place as that assembled by others in different times and places.

Taken together, we see that the philosophical anchoring of epistemology in experience, truth in the service of goals, and consensus in the conjoining of embodied views suggest that knowledge reflects the negotiation between the individual, society, and the world. This negotiation can result in several relationships, ranging from passive acceptance from authority, unstinting adherence to a particular perspective because it has worked in the past, and, finally, the active and open consideration of the meaning of past and current interactions with the world and with other persons as the basis of knowledge.

EXAMPLES OF AN EPISTEMOLOGY OF CONNECTION

The examples presented here represent, for the most part, programs that demonstrate a clear understanding that knowledge always has a component that connects the knowers individually and as a group to the knowledge that is taught. In the next section, we will consider psychological connections of the individual learner to knowledge, something that is different and much more common. Here, though, we seek cases where a curriculum or instructional system is clearly developed with an acknowledgment of the connection.

The opening example from the *Meno* raises the question of the status of Socratic instruction in chemistry. Often, Socratic teaching is reduced to various forms of teaching through questions (Chin 2007). However, there are also examples where the questions constitute a dialogue between an expert and a novice while they consider the evidence of a particular experience (which can be virtual, actual, or paper based). For example, in chemistry, Holme (1992) has reported on using spontaneous groups as the basis of Socratic questioning in lecture, but he also starts the session off with a simple shared experience, such as turning the lights on and off to form the basis of consideration of what occurs during the flow of electricity. DePierro et al. (2003) present a more detailed set of examples, with specific reference to student experience as a basis of discussion. Finally, Spencer and Lowe (2003) use a Socratic dialogue between an experienced teacher and a new teacher to discuss important questions of entropy, making particular use of the format as a way of exchanging information in a shared inquiry. This highlights that Socratic teaching can also be seen as the basis of guided inquiry instruction, now well represented by the Process Oriented Guided Inquiry Learning (POGIL) initiative that Spencer helped to found (Farrell et al. 1999; Spencer 1999).

Pragmatist philosophy, at least as formulated by James, Dewey, Peirce, and others, is essentially absent from chemistry instruction as an explicit viewpoint. Its cognitive connections are present, as discussed in the next section. But pragmatist views of knowledge are, I would argue, the basis of certain strands of constructivism, including that developed by Ernst van Glaserfeld (Tobin 2007), and were brought into chemistry education by George Bodner (1986; Bodner et al. 2001). Although he is also very clear that he adheres to the personal constructivism that is the basis of cognitive connections, Bodner does not shy away from following van Glaserfeld's arguments, stating, for example, "From the perspective of the constructivist and radical constructivist theories, knowledge should no longer be judged in terms of whether it is true or false, but in terms of whether it works. The only thing that matters is whether the knowledge we construct functions satisfactorily in the context in which it arises." Bodner then relates this position to a variety of corollary ideas in teaching, including those specifically suggested from the work of Jean Piaget and George Kelly.

Bodner's rich pedagogical work bears the mark of this philosophical stance, but there are few examples where the philosophical basis is presented plainly to students. One reason, we can infer, is that students may have to develop a more traditional view of a field before understanding the idea that the knowledge they are developing is itself always connected to the knowers who developed it. Working originally from Bodner's perspective, Bhattacharyya (2008) studied the epistemic development of organic chemists through a cross-age study of undergraduate students, graduate students, and "seasoned" practitioners. Throughout, a functional approach to knowledge was found with the specific experiences of the learner, in the classroom and the research laboratory, shaping the conceptual development. Even the deep conceptual understanding of the seasoned practitioners was linked to the formation of a professional identity.

The feminist movement in philosophy is also matched well by ideas developed for a feminist pedagogy (Middlecamp and Subramaniam 1999). Part of this movement emphasizes the goal of giving students a voice in science, as discussed in the sociology of connection. But the direct consideration of epistemological considerations has been used also, when "Feminist classrooms explore the origins of ideas and theories, the *position* of those who put them forth, and the factors that influence how knowledge came to exist in its present forms." Doing this in a thoroughgoing way is rare in chemistry education or in science education generally. Part of the reason is a tension inherent in teaching content for students to master while they also learn the situated origin of that knowledge (Richmond et al. 1998).

Finally, incorporating the history, philosophy, and sociology of science into chemistry curricula aligns instruction with Standard G of the National Science Education Standards, which includes the idea that science is a human endeavor (Rasmussen et al. 2008, but see also Erduran and Scerri 2002 and Scerri 2003 for other viewpoints). There is a risk that oversimplification leads to myth making as the details of history and sociology are trimmed to tell the direct chemical story. But there are also very good examples of specific and rich instruction, such as the story of the creation, use, and banning of chlorofluorocarbons. In this way, a variety of facts, including facts about chemical and physical properties, change the meaning of chlorofluorocarbons over time.

THE COGNITION OF CONNECTION

The second reason for considering connection as a basis of learning is its relation to the ways in which the mind works in itself. This brings in questions of cognition, ranging from questions of what we do with sensory information to how we interpret sense phenomena, consciously and unconsciously.

In this case, too, we have deep historical roots to contemporary ideas. The most prominent of these is in the writings of John Dewey. Dewey's work can be classified into many different disciplines, including education, psychology, philosophy, and political science. Primarily, he was a theoretical thinker, but he always linked his work to practical problems, including those of democracy. He was a member of the pragmatist movement, linking with many of the ideas of James.

His emergence as a major thinker was solidified with the paper "The Reflex Arc Concept in Psychology" (Dewey 1896), where he criticizes the idea of analyzing psychological experiments in terms of "a patchwork of disjointed parts," including sensation, ideas, and the nerve action that constitutes a response. Instead, he argued for analyzing stimulus, cognition, and action as a single whole, where sensation and reaction are coordinated, not sequential. A consequence of this is that sensation becomes deliberate, in response to and in anticipation of information from the outside world. Furthermore, what information comes in from the outside is constrained by prior experience, for prior experience provides the guide to interpreting that experience. Even in cases where sensations are instinctive, Dewey points, they are adaptive to the environment in which the organism typically exists. This creates a continuous circuit that, in Dewey's analysis, is used repeatedly so long as it creates no problems. When problems occur—for example, when a child reaches for an interesting object to grasp it but finds a burn instead—refinement of the circuit occurs to include new ideas, new sensations, and new response acts. Throughout, it is experience that provides the basis for action, but it is thought (what he calls "psychic response") that allows for reconstruction of a disrupted circuit.

Dewey moved his ideas quickly into questions of practice for education, including the work that he and his wife did in founding a laboratory school that focused on connections of thought with experience in the determination of action, as the basis of instruction. He recognized that this was quite radical, as he indicates in an address to the Psychological Society in 1900, where the heart of Deweyan psychology—experience, thought, and response existing in a circuit—therefore became a goal for pedagogy:

With the adult we unquestioningly assume that an attitude of personal inquiry, based upon the possession of a problem which interests and absorbs, is a necessary precondition of mental growth. With the child we assume that the precondition is rather the willing disposition which makes him ready to submit to any problem and material presented from without. Alertness is our ideal in one case; docility in the other. With one, we assume that power of attention develops in dealing with problems which make a personal appeal, and through personal responsibility for determining what is relevant. With the other we provide next to no opportunities for the evolution or problems out of immediate experience, and allow next to no free mental play for selecting, assorting and adapting the experiences and ideas that make for their solution.

Dewey, who was as much a political scientist and philosopher as he was a psychologist and educator, did little concrete work to link his assertions to experimental data of his own, even within the schools that he started or supported. Providing a psychological rationale (or theory) to explain why Dewey's connected learning works fell to others. Important in this are Vygotsky, who worked on language acquisition, and Piaget, who introduced the idea of the role of personal knowledge structures that need to be deliberately challenged for learning to occur. Recent work has looked more carefully at the developmental psychology that explains why learning and doing science in context may be so effective. Throughout, a persistent theme is that the person, as the place where knowledge is constructed and maintained, will bring his or her own reasons to the knowledge.

The most recent scholarship in learning science through cognitive connective strategies has occurred in work on *situated cognition* (Brown et al. 1989). The idea of concepts as tools is central to the idea, leading to the idea of learning as *cognitive apprenticeship*. From this, several characteristics of situated learning emerge, as presented for the specific context of math learning (Brown et al. 1989):

- By beginning with a task embedded in a familiar activity, it (cognitive apprenticeship) shows the students the legitimacy of their implicit knowledge and its availability as scaffolding in apparently unfamiliar tasks.
- By pointing to different decompositions, it stresses that heuristics are not absolute but are assessed with respect to a particular task, and that even algorithms can be assessed in this way.
- By allowing students to generate their own solution paths, it helps make them conscious, creative members of the culture of problem-

solving mathematicians. And, in enculturating through this activity, they acquire some of the culture's tools—a shared vocabulary and the means to discuss, reflect upon, evaluate, and validate community procedures in a collaborative process.

Of course, learning is often explicitly associated with apprenticeship to a trade or profession, and it is easy to design learning environments that fit the idea of apprenticing students for inclusion in chemistry. But the idea of connecting through cognition also has potential meeting in situations in which the student is not planning on being a chemist (or an engineer, or a doctor). There, the situation becomes a meaningful context for learning how new knowledge relates to the students in a potential, not in an actual, way.

The idea of situated cognition also relates to learning to use analogs of other domains. For example, anthropomorphic metaphors are introduced to provide students with a way of describing phenomena in psychological terms. Recently, the developmental psychologist Alison Gopnik has formulated the idea that cognition in science recapitulates the cognition that *every* child uses in learning. She emphasizes that every child has the challenge of learning about three things: language, other persons, and the world. And for all three, her evidence suggests, active theory formation and evaluation, accompanied by testing of theories through specific and intentional experiments, are the basis of how we learn how to talk, how to manage objects in the world, and how to work in an intensely social environment. While she underscores that science carries the study of "objects in the world" to another level while using the same cognitive skills, it is also important to recognize the importance of learning about "other minds" to the process of learning. If we are indeed skilled in figuring out, reacting to, and, if needed, manipulating, other minds, then the learning that connects content to human behavior may be effective. This can occur in two ways. First of all, since we are good at describing what other people do in terms of volition, ascribing volition to natural systems may make it easier to understand how they work, though of course, only in an analogical way. This is found prominently in anthropomorphic language and scenarios in learning. The second way in which our psychological impulse may be useful in learning may occur when we can connect learning to existing social systems. We make learning a component of working well in the social world, for example, through group learning and through role playing that allow us to bring our interest in working with and controlling people into how we learn chemistry.

EXAMPLES OF CONNECTION THROUGH COGNITION

It is probably a trivial statement to say that learning has to involve the cognition of the learner. However, in many cases, the curriculum or other aspects of instruction are passive in that regard, providing knowledge as a set of standard items to be acquired and used by students following a view of learning as a transmission process. Connections through cognition occurs when instruction explicitly uses the learner's capacity to make a personal meaning part of the cognitive process. In this case, many examples are available in chemistry. Four different strategies are available:

- problem-based learning, including role modeling;
- inquiry learning;
- metacognitive strategies incorporating students' thinking about themselves and their society; and
- use of analogies from other domains, including anthropomorphism.

I have already discussed how the first of these is present in chemistry education (Wink 2005) when instruction aims at making learning "relevant" to students. Problem-based learning fits well within the concept of situated cognition. Since problem-based learning requires students to use their knowledge to accomplish a task, it fits also in the category of "design" inquiry (Rudolph 2005). Design inquiry seeks to make knowledge connect to a goal that may lie outside the traditional curriculum, but it has the advantage of both bringing outside motivation (whether real or simulated) into the learning process, thereby connecting students with the "real" world. This can be carried out in limited ways in specific laboratories as in the general chemistry program *Working with Chemistry* (Wink et al. 2005), during unit-long exercises as in the high school *Chemistry in the Community* project (American Chemical Society [ACS] 2008), or over an extended period, as Gallagher-Bolos and Smithenry (2004) show in a year-long high school curriculum where students form a community simulating a soap company.

Inquiry, as Rudolph (2005) notes, is also something that occurs within a discipline and as students learn the content of the discipline. In this case, it fits more with definitions of inquiry that have students investigate concepts and processes within a subject, as suggested by Abraham (2005) and as used to assess curricula by Bretz, Towns, and their coworkers (Fay et al. 2007; Buck et al. 2008). In that case, inquiry is primarily characterized by the extent to which the problem, the procedure, or both were developed by the student. The cognitive connection here is found when the student must generate important elements of the experiment, presumably through thinking that connects with the students' prior knowledge and plans for the experiment.

Metacognition, where students are aware of and regulate their thinking as part of the learning process, is a recent addition to the chemistry education toolkit. Rickey and Stacy (2000) discuss different examples of this, including the specific example of having students discuss their work to share and critique their ideas in a common project. In chemistry, strategies include paper- and computer-based procedures to document students' conceptual reasoning through tools such as thinking frames (Mattox et al. 2006), concept maps (Francisco et al. 2002), cooperative groups (Cooper et al. 2008a), and work in using extended writing within laboratory work (Greenbowe and Hand 2005). Recently, Cooper and coworkers (2008a,b) have studied this to uncover to what extent a students' metacognition affects complex problem solving and, more importantly, how instruction with metacognition can prompt some learners to shift their problem-solving strategies from less to more productive modes.

Perhaps the most general way in which metacognition can be used to connect students to the learning of chemistry is through writing that includes a deliberate reflective component. This has been well documented in science and in chemistry through work on the science writing heuristic (SWH) (Hand 2007), part of the general idea of writing to learn in science (Wallace et al. 2007). Such strategies undoubtedly aid in students' learning to understand and to use the complex vocabulary of science (Wellington and Osborne 2001). Writing is also a means to support knowledge growth (Keys 1999). The SWH allows students to connect through the acts of developing beginning questions, designing and executing a data collection plan to answer the question, developing claims based on evidence, and reflecting on their experience and its relation to knowledge. As a result, when implemented well, the SWH benefits learning of basic chemical concepts, including, for example, thermochemistry (Rudd et al. 2001). At several different points, students have to formulate their own responses, from the initial questions to the final reflection, providing strong connections between thinking and activity.

The cognitive connections available through analogies are much more widespread in part because analogies are probably a basic cognitive skill themselves. Orgill and Bodner (2005) discuss that analogies allow instructors to support student growth in a new domain (like bio-chemistry) through connections to prior knowledge in a familiar domain. Scores of articles have been published on the use of analogies. Some reflect the fundamental analogies used to describe chemical and physical systems at the atomic–molecular scale using macroscopic analogies, such as standing waves (Davis 2007). Others provide students with graphic or analogical organizers for their problem solving (Ault 2006). Also, the use of any macroscopic model is an example of analogy (Justi and Gilbert 2002). Many introduce human sensory constructs into the analysis of the actions of molecules (e.g., sight: Mattice 2008). Problems arise, however, when students either do not appreciate the analogy because they do not understand the analog domain or, in other cases, elements from the analog domain can transfer into the target domain, causing confusion and misconceptions.

Perhaps the most interesting kind of analog connection is in ascribing psychological states to atoms and molecules. This is embedded in the basic classificatory schemes of chemistry, as in the concepts of hydrophilicity ("loving water") and hydrophobicity ("fearing water") in describing the interactions of substances with water. Such psychologizing and the associated anthropomorphic language permits instructors to take on the wide range of experience that the students have with people and to use it in the classroom, perhaps exploiting a basic human cognitive function in solving the problem of "other minds," which is so vital for a social animal (Gopnik). Certainly explicit anthropomorphism is widespread in chemistry education, beautifully illustrated by the work of Primo Levi in *The Periodic Table*, where the analogy between a person or a group and an element is used as the driver for his writing. Having students create their own analogies is a form of metacognition, but also runs the risk of stereotyping "people as well as molecules" in dangerous ways (Miller 1992). In addition, carrying things too far with any analogy occurs in anthropomorphism also (Taber and Watts 1996). Certainly, this is the case when the analogies move into the area of sexual attraction, both explicitly and implicitly (Biology and Gender Study Group 1989), that describes chemical reactions between small energetic nucleophiles and large, passive substrates in violent terms of attack.

THE SOCIOLOGY OF CONNECTION

Our last perspective comes from considering the relationship of connections between scientific knowledge and learning on the one hand and society on the other. This is a reciprocal relationship, where we consider how what is known in science and how science is taught both affect society and, conversely, how society affects what is known in science and how science is taught. As with epistemology and cognition, the role of sociology in science is generally acknowledged by all at a trivial level: It is true that humans in society carry out science, teach science, and learn science. Hence, science always has a human element, but traditional views of science probably emphasize that science itself ultimately transcends these social elements (Haack 1996) and that learning ultimately is about participating in a set body of knowledge, not participating in its construction (Matthews 1994; Scerri 2003).

Instrumentalist views of science challenge strongly these views of the subject. Steve Fuller, a sociologist and philosopher, has, for example, cast these differences in the form of a contrast between "disenchanted science" and "enchanted science," contrasted in Table 1.1 below.

The consequence of the first kind of science, he argues, is a public one and the alienation of students from science because, in almost all areas of life, knowledge has moral, value, and political components. In this case, therefore, learning that connects with these kinds of "enchantment" is likely to be considered more legitimate because it invites the learner to consider how participation in science extends from his or her normal activities in society. Furthermore, Fuller elsewhere emphasizes that science has a "governance" structure that reflects those who will use (or abuse) science. He sees significant problems that occur when we avoid this governance structure, which arises when science becomes a special subject. Instead, Fuller argues for embracing the structure and for acknowledging the sociological facts that all science reflects society, including the society of those who are attempting to learn it.

Fuller and others in the science studies movement such as Bruno Latour (1999) and Helen Longino (2002) are primarily analytic in his approach, working to develop and to interpret evidence of the enduring connection science content and learning to society. Another thread in

TABLE 1.1 Differences between Disenchanted Science and Enchanted Science

Disenchanted Science	Enchanted Science
… is a value-neutral mode of inquiry.	… sees reality with a moral center.
… has norms that summarize empirical phenomena.	… has norms that are value laden.
… sees the "nature of consciousness" and the "meaning of life" as mere pseudophenomena.	… posits effects judged by clarification of ends and activation of political will.

the sociology of connection considers the issue of participation (or nonparticipation) of different groups in science. This reflects pedagogical questions raised by thinkers like Paolo Friere and bell hooks, who argue for teaching that specifically targets the problems of those who are not participants as the basis of what Friere, in one of his most famous texts, calls the "pedagogy of the oppressed." In this case, the solution to a science that excludes, or seems to exclude, other groups is to take those lives and to make them the basis of science and science learning. This is both a correction to the distortion of the current system and a way to bring other groups into learning.

A particularly effective discussion of this in the context of learning science was presented by Maralee Mayberry (1996), also a sociologist, in her analysis of "reproductive and resistant pedagogies." In analyzing conventional views of collaborative learning, she finds that it "engages students in an ongoing conversation about these aspects of disciplinary knowledge, and in turn trains students who are capable of adapting to the 'conventions of conversation' in educated communities." This, though, occurs within a sociological framework of bringing students into the viewpoints of the *unchallenged* community that exists before learning. Hence, she sees reproductive pedagogy as maintaining the sociological relationships of insider and outsider. In contrast, she sees that "the feminist classroom conversation and process of knowledge production are overtly political and aimed toward social and educational change," that if we want to have a change in science to make it more responsive to social needs, we need to bring in both a critique and a participation scheme to allow students to raise questions about what goes on now.

EXAMPLES OF CONNECTION THROUGH SOCIOLOGY

Connections through sociology can occur in three different ways. These follow a schema suggested by Capobianco (2007), working from a perspective of feminist pedagogy. One examines the classroom as a place where inclusion needs to be fostered by bringing new audiences to the learning of science. The second way for a connection through sociology addresses changes in the categories of a field so that the field itself encompasses more areas of inquiry, including other groups that have been traditionally excluded. The third looks at the actual practices of a field and attempts to engage in "transformative practices." These are similar to a schema used to survey the literature on girls and science (Brotman and Moore 2008).

Inclusion of more groups in a field can be as simple as providing instruction specifically to those who are not usually taught a field. This is present, for example, in many different programs that target a group or groups for teaching and is the basis of programs for girls and women (e.g., Jayaratne et al. 2003; Brotman and Moore 2008). In this case, the step of bringing a target population into contact with science is supposed to have an effect of enhancing understanding and of increasing interest. In some cases, specific content that may be relevant to the target population is included, for example, in the Summer Science for Girls program that provided eighth-grade girls with hands-on instruction in science while also exposing the participants to women scientists as role models and providing "exercises aimed at dispelling the stereotypes associated with women doing science." Such a combination of content and motivational work was also present in a chemistry-based outreach program from Simmons College (Lee and Schreiber 1999).

Targeted interventions as a means for connecting chemistry to new groups are also present implicitly in many programs that work with the educational venues where many students take chemistry. Prominent among these are efforts to affect chemistry instruction at community colleges, the site of the majority of undergraduate general chemistry courses. This is prominent as a driving force behind the recent work of the National Science Foundation Division of Undergraduate Education, which has allowed for expanded funding of Course, Curriculum, and Laboratory Improvement projects that include community college participation. Perhaps more important is the concept that community colleges are themselves a source of valuable insight into how to include more students in chemistry learning, for in those environments, instruction is much more likely to be responsive to the particular sociological makeup of a population, as recently highlighted by the ACS in a discussion of Hispanic-serving institutions (ACS 2008). The impact of such inclusionary practices as the ACS Scholars and Project Seed programs is good evidence that connection by simply exposing more students to chemistry, especially research chemistry, is in itself an effective means for increasing participation.

Although inclusion is a necessary step in connecting more students to chemistry, there is also good reason to suspect that additional steps are needed. Thus, the second way in which a connection through sociology can occur is by incorporating specific discussions of race, ethnicity, and gender within instruction. The theoretical and methodological components of such cross-cultural instruction, including implications for research and practice on student connections, have been explored by Aikenhead and Jegede (1999), where they examined the ways in which instruction involves "cultural border crossing" that may either exclude or include students. In this case, factors such as students' home lives, the ability to "play in the new environment," and resistance as "science teaching ... attempt[s] to assimilate them" all affect student movement into science.

An alternative to assimilation is to make clear to students that they may cross the sociological border into science with their cultural identity both intact and, perhaps, valued. This is a specific goal of the work of Middlecamp and Subramaniam (1999). Within the biological sciences, including gender as a specific component of course content is a more obvious move than in chemistry (Birke 2001). There are more studies in which a connection of chemistry to sociology occurs through the lens of race and ethnicity, for example, in the Project Inclusion work of Hayes and Perez (1997) and the multicultural examples developed by Middlecamp and Fernandez (1999).

The third way in which connection on the basis of sociology can develop is when educators and students engage in learning that has the potential to change the way in which science works in itself. As noted earlier, such a transformation is easier to contemplate and perhaps to accomplish in areas where social interactions seem to be more prominent, for example in biology. Considerably less is known about how science might change if, for example, women's voices were more prominent. At the very least, the relationship of the history of the physical sciences and military goals has been well documented (Schiebinger 1999; Fuller 2002; Rudolph 2002). But the specific case of chemistry— whether chemistry as traditionally taught reflects the perspectives of a limited sociological group—is not well studied. Some work, noted earlier, examined how images of chemical reactions as a site for a passive (i.e., feminine) receptor molecule undergoing attack by an energetic nucleophile may reflect a gendered history within the description and understanding of chemical reactions (Biology and Gender Study Group 1989).

Little is known about how chemistry might change if the influence of other sociological groups (women, persons of color, non-Western cultures) were given more control. An exception is present in the work of Aikenhead and Ogawa (2007). They documented how three different world views (North American indigenous, Japanese, and Euro-American) approach not only knowledge but also what is known. In particular, they recognize a fundamental difference between the views that science is a distinct way of knowing (in Euro-American traditions) compared to efforts to make science a part of traditional knowledge cultures. The emergence of "ethnoscience" within environmental science and biology exemplifies that specific knowledge and ways of thinking from traditional groups may indeed influence how the world is viewed (Snively and Corsiglia 2000).

These previous studies, because they are based outside of chemistry, do not address the question of how chemistry itself might be transformed if standpoints other than those of white males were included as a fundamental part of chemistry knowledge. This does not mean that connecting different social groups to the production of knowledge will not change knowledge; rather, it is a suggestion that this is an area that still needs to be researched, probably in conjunction with teaching of nontraditional groups. One set of perspectives on how this may occur comes as a consequence of Sandra Harding's (2001) discussion of why feminist standpoint epistemology suggests new directions for science. Among her "'grounds' for the feminist claims" are the following:

- The neglect of women's lives as starting points for research implies that including these may lead to new avenues of research.
- Considering the "strangeness" of women to the social order in science prompts researchers to view the dominant viewpoints as possibly defective while the traditional oppression experienced by women means they have an interest in keeping a focus on examining the basis of knowledge. Together, these mean that women's standpoints are the right starting point to examine whether the traditional knowledge is valid.
- Women have played a very large role in the production of "craft" knowledge over time. Considering craft knowledge as a starting point for science has the potential to create new descriptions and rationales for science, and therefore perhaps new kinds of science.

These points can easily be changed to include other nontraditional social groups and, as noted, there are examples of connecting the learning of chemistry to the lives of other cultures. But these are few, and new ideas and research in this area is necessary before the questions of "what would a nontraditional chemistry look like?" can be answered.

REFERENCES

Abraham, M. R. (2005) Inquiry and the learning cycle approach. In *Chemists Guide to Effective Teaching*, eds. N. Pienta, M. M. Cooper, and T. Greenbowe. Upper Saddle River, NJ: Prentice Hall.

Aikenhead, G. S. and O. J. Jegede (1999) Cross-cultural science education: A cognitive explanation of a cultural phenomenon. *J. Res. Sci. Teach. 36,* 269–287.

Aikenhead, G. S. and M. Ogawa (2007) Indigenous knowledge and science revisited. *Cult. Stud. Sci. Educ. 2,* 539–560.

American Chemical Society (ACS) (2008) *Workshop on Increasing Participation of Hispanic Undergraduates in Chemistry.* Washington, DC: American Chemical Society.

Ault, A. (2006) Mole city: A stoichiometric analogy. *J. Chem. Educ. 83,* 1587–1588.

Bhattacharyya, G. (2008) Who am I? What am I doing here? Professional identity and the epistemic development of organic chemists. *Chem. Educ. Res. Pract. 9,* 84–92.

Biology and Gender Study Group (1989) The importance of the feminist critique for contemporary cell biology. In *Feminism & Science,* ed. N. Tuana. Bloomington, IN: Indiana University Press.

Birke, L. (2001) In pursuit of difference: Scientific studies of women and men. In *The Gender and Science Reader,* eds. M. Lederman and I. Bartsch. New York: Routledge.

Bodner, G. (1986) Constructivism: A theory of knowledge. *J. Chem. Educ. 63,* 873–878.

Bodner, G., M. Klobuchar, and D. Geelen (2001) The many forms of constructivism. *J. Chem. Educ. 78,* 1107.

Brickhouse, N. W. (2001) Embodying science: A feminist perspective on learning. *J. Res. Sci. Teach. 38,* 282–295.

Brotman, J. S. and F. M. Moore (2008) Girls and science: A review of four themes in the science education literature. *J. Res. Sci. Teach. 45,* 971–1002.

Brown, J. S., A. Collins, and P. Deguid (1989) Situated cognition and the culture of learning. *Educ. Res.* Jan.-Feb., 32–42.

Buck, L. B., S. L. Bretz, and M. H. Towns (2008) Characterizing the level of inquiry in the undergraduate laboratory. *J. Coll. Sci. Teach.* Sept.–Oct., 52–58.

Capobianco, B. M. (2007) Science teachers' attempts at integrating feminist pedagogy through collaborative action research. *J. Res. Sci. Teach. 44,* 1–32.

Chin, C. (2007) Teacher questioning in science classrooms: Approaches that stimulate productive thinking. *J. Res. Sci. Teach. 44,* 815–843.

Cooper, M. M., S. Sandi-Urena, and R. Stevens (2008a) Reliable multi-method assessment of metacognition use in chemistry problem-solving. *Chem. Educ. Res. Pract. 9,* 18–24.

Cooper, M. M., C. T. Cox, M. Nammouz, E. Case, and R. Stevens (2008b) An assessment of the effect of collaborative groups on students' problem solving strategies and abilities. *J. Chem. Educ. 85,* 866–872.

Davis, M. (2007) Guitar strings as standing waves: A demonstration. *J. Chem. Educ. 84*, 1287–1289.

DePierro, E., F. Garafolo, and R. T. Toomey (2003) Using a Socratic dialog to help students construct fundamental concepts. *J. Chem. Educ. 80*, 1408–1416.

Dewey, J. (1896) The reflex arc in psychology. *Psychological Review III*, July, 357–370.

Dewey, J. (1900) Psychology and social practice *Psychological Review VII*, March, 105–124.

Dewey, J. (1910) Science as subject matter and as method. *Science 331*, 121–127.

Erduran, S. and E. Scerri (2002) The nature of chemical knowledge and chemical education. In *Chemical Education: Towards Research-Based Practice*, eds. J. K. Gilbert, O. De Jong, R. Justi, D. F. Treagust, and J. H. van Driel. Dordrecht: Kluwer.

Farrell, J. J., R. S. Moog, and J. N. Spencer (1999), A guided inquiry general chemistry course. *J. Chem. Educ. 76*, 570–574.

Fay, M. E., N. P. Grove, M. H. Towns, and S.L. Bretz (2007). A rubric to characterize inquiry in the undergraduate chemistry laboratory. *Chem. Educ. Res. Pract. 8*, 212–219.

Francisco, J. S., M. B. Nakhleh, S. C. Nurrenbern, and M. L. Miller (2002) Assessing student understanding of general chemistry with concept mapping. *J. Chem. Educ. 79*, 248–257.

Fuller, S. (2002) *Thomas Kuhn: A Philosophical History for Our Times*. Chicago: University of Chicago Press.

Gallagher-Bolos, J. A. and D. W. Smithenry (2004) *Teaching Inquiry-Based Chemistry*. Portsmouth, NH: Heinemann.

Greenbowe, T. J., and B. Hand (2005) Introduction to the science writing heuristic. In *Chemists' Guide to Effective Teaching*, eds. N. J. Pienta, M. M. Cooper, and T. J. Greenbowe, pp. 140–154. Upper Saddle River, NJ: Prentice Hall.

Haack, S. (1996) Science as social? Yes and no. In *Feminism, Science, and the Philosophy of Science*, eds. L. H. Nelson and J. Nelson, pp. 79–93. Dordrecht: Kluwer.

Hand, B. (2007) *Science Inquiry, Argument, and Language: The Case for the Science Writing Heuristic*. New York: Sense Publishers.

Haraway, D. J. (1988) Situated knowledges: The science question in feminism and the privilege of partial perspective. *Fem. Stud. 14*, 575–600.

Harding, S. (2001) Feminist standpoint epistemology. In *The Gender and Science Reader*, eds. M. Lederman and I. Bartsch. New York: Routledge.

Harding, S. (2004) Introduction: Standpoint theory as a site of political, philosophic, and scientific debate. In *The Feminist Standpoint Theory Reader*, ed. S. Harding. New York: Routledge.

Hayes, J. and P. Perez (1997) Project inclusion: Native American plant dyes. *Chem. Herit. 15*(1), 38–40.

Holme, T. A. (1992) Using the Socratic method in large lecture classes. *J. Chem. Educ. 69*, 974–977.

James, W. (2000) *Pragmatism and Other Writings*. New York: Penguin Putnam.

Jayaratne, T. E., N. G. Thomas, and M. Trautmann (2003) Intervention program to keep girls in the science pipeline: Outcome differences by ethnic status. *J. Res. Sci. Teach. 40*, 393–414.

Justi, R. and J. K. Gilbert (2002) Models and modeling in chemical education. In *Chemical Education: Towards Research-Based Practice*, eds. J. K. Gilbert, O. De Jong, R. Justi, D. F. Treagust, and J. H. van Driel. Dordrecht: Kluwer.

Keys, C. W. (1999) Revitalizing instruction in scientific genres: Connecting knowledge production with writing to learn in science. *Sci. Educ. 83* 115–130.

Latour, B. (1999) *Pandora's Hope: Essays on the Reality of Science Studies*. Cambridge, MA: Harvard University Press.

Lee, N. E. and K. J. Schreiber (1999) The Chemistry Outreach Program: Women undergraduates presenting chemistry to middle school students. *J. Chem. Educ. 76*, 917–918.

Longino, H. E. (2002) *The Fate of Knowledge*. Princeton, NJ: Princeton University Press.

Matthews, M. R. (1994) *Science Teaching: The Role of History and Philosophy of Science*. Routledge: New York

Mattice, J. (2008) If you were a molecule in a chromatography column, What would you see? *J. Chem. Educ.*,

Mattox, A. C., B. A. Reisner, and D. Rickey (2006) What happens when chemical compounds are added to water? An introduction to the model-observe-reflect-explain (MORE) thinking frame. *J. Chem. Educ. 83*, 622–624.

Mayberry, M. (1996) Reproductive and resistant pedagogies. *J. Res. Sci. Teach. 35*, 443–459.

Middlecamp, C. H. and M. A. D. Fernandez (1999) From San Juan to Madison: Cultural perspectives on teaching general chemistry. *J. Chem. Educ. 76*, 388–391.

Middlecamp, C. H. and B. Subramaniam (1999) What is feminist pedagogy? Useful ideas for teaching chemistry, *J. Chem. Educ. 76* 520–525.

Miller, L. L. (1992) Molecular anthropomorphism: A creative writing exercise. *J. Chem. Educ. 69*, 141–142.

National Research Council (2000) *How People Learn: Brain, Mind, Experience, and School: Expanded Edition*. Washington, DC: National Academy Press.

National Research Council (2005) *Inquiry and the National Science Education Standards*. Washington, DC: National Academy Press.

Orgill, M., and G. M. Bodner (2005) The role of analogies in chemistry teaching. In *The Chemists' Guide to Effective Teaching*, eds. T. G. Greenbowe, N. Pienta, and M. M. Cooper, pp. 90–95. Upper Saddle River, NJ: Pearson.

Plato (1961) Meno *(Tr. W. K. C. Guthrie)*. In *Plato: Collected Dialogs*, eds. E. Hamilton and H. Cairns. Princeton, NJ: Princeton University Press.

Rasmussen, S. C., C. Giunta, and M. R. Tomchuk (2008) Content standards for the history and nature of science. In *Chemistry and the National Science Education Standards*, 2nd ed., ed. S. L. Bretz. Washington, DC: American Chemical Society.

Richmond, G., E. Howes, L. Kurth, and C. Hazelwood (1998) Connections and critique: Feminist pedagogy and science teacher education. *J. Res. Sci. Teach. 35*, 897–918.

Rickey, D. and A. Stacy (2000) The role of metacognition in learning chemistry. *J. Chem. Educ. 77*, 915–920.

Rudd, J. A. III, T. J. Greenbowe, B. M. Hand, and J. J. Legg (2001) Using the science writing heuristic to move towards an inquiry-based laboratory curriculum: An example from physical equilibrium. *J. Chem. Educ. 78*, 1680–1686.

Rudolph, J. L. (2002) *Scientists in the Classroom*. New York: Palgrave.

Rudolph, J. L. (2005) Inquiry, instrumentalism, and the public understanding of science. *Sci. Educ. 89*, 803–821.

Scerri, E. R. (2003) Philosophical confusion in chemical education research. *J. Chem. Educ. 80*, 468–477.

Schiebinger, L. (1999) *Has Feminism Changed Science?* Cambridge, MA: Harvard University Press.

Scott, D. (2007) *Recollection and Experience: Plato's Theory of Learning and Its Successors*. Cambridge: Cambridge University Press.

Snively, G. and J. Corsiglia (2000) Discovering indigenous science: Implication for science education. *Sci. Educ. 85*, 6–34.

Spencer, J. N. (1999) New directions in teaching chemistry: A philosophical and pedagogical basis. *J. Chem. Educ. 76*, 566–569.

Spencer, J. N. and J. P. Lowe (2003) Entropy: The effect of distinguishability. *J. Chem. Educ. 80*, 1417–1424.

Taber, K. S. and M. Watts (1996) The secret life of the chemical bond: Students' anthropomorphic and animistic references to bonding. *Int. J. Sci. Educ. 18*(5), 557–568.

Tobin, K. (2007) Key contributors: Ernst van Glaserfeld's radical constructivism. *Cult. Sci. Educ. 2*, 529–538.

Wallace, C. S., B. Hand, and V. Prain (2007) *Writing to Learn in the Science Classroom*. London: Springer.

Wellington, J. and J. Osborne (2001) *Language and Literacy in Science Education*. Philadelphia, PA: Open University Press.

Wink, D. J. (2005) Relevance and learning theory. In *Chemists Guide to Effective Teaching*, eds. N. Pienta, M. M. Cooper, and T. Greenbowe. Upper Saddle River, NJ: Prentice Hall.

Wink, D. J., S. F. Gislason, and J. E. Kuehn (2005) *Working with Chemistry*, New York: W. H. Freeman.

CHAPTER 2

Chemistry and the Environment: A SENCER Model Course

AMY M. SHACHTER

Santa Clara University, Santa Clara, CA

INTRODUCTION

Science Education for New Civic Engagements and Responsibilities[1] is a science education reform project supported by the National Science Foundation through the National Center for Science and Civic Engagement. SENCER is intended to help improve science education within a broad framework of connected learning by teaching science principles through public issues. In SENCER courses, students actively participate in civic engagement as a means of learning science. This chapter will focus on a SENCER model course, Chemistry and the Environment.[2,3] The course is primarily for nonscience majors and typically has an enrollment ranging from 25 to 48 undergraduates. Chemistry and the Environment exemplifies the SENCER ideals by connecting basic chemical principles to complex, public issues through campus-

---

[1] Information about the SENCER project can be found at http://www.sencer.net (accessed January 2009).

[2] Previously described in Shachter, A. and J. Edgerly (1999) Campus environmental resource assessment projects for non-science majors. *J. Chem. Educ.* 76, 1667–1670.

[3] Briefly described in Middlecamp, C., T. Jordan, A. Shachter, S. Lottridge, and K. Oates (2006) Chemistry, society, and civic engagement (part 1): The SENCER project. *J. Chem. Educ.* 76, 1301–1307.

Any opinions, findings, and conclusions or recommendations expressed in this material are those of the author(s) and do not necessarily reflect the views of the National Science Foundation.

---

and community-based research projects as well as course content and structure. The research projects provide a mechanism for students to learn chemistry fundamentals and to sharpen their understanding of and appreciation for the process, contributions, and limitations of science. With a campus or community focus, the projects are designed to be "local," connecting to issues relevant to the daily lives of the students. The local projects inspire students to view themselves as community stakeholders and, consequently, foster civic engagement and responsibility beyond the course. All aspects of the course from learning outcomes to examples of campus engagement are presented in this chapter.

SENCER BACKGROUND

Initiated in 2001, SENCER has established and supported an ever-growing community of faculty, students, academic leaders, and others to improve undergraduate science, technology, engineering, and mathematics (STEM) education for nonscience majors by connecting learning to critical civic questions. The SENCER community counts as its members over 1100 educators, academic leaders, and students from over 300 colleges and universities from the United States and 13 other countries. SENCER is housed in the National Center for Science and Civic Engagement, which was established in affiliation with Harrisburg University of Science and Technology.

In 2007, the SENCER project expanded to establish five regional SENCER Centers of Innovation (SCIs). The SCIs expand the work of SENCER by organizing regional workshops designed to foster a multi- and interdisciplinary approach to science education with a focus on civic engagement. The SCIs create a multi-institutional community with outreach to regional academic institutions and related organizations as well as to strategic national institutions and organizations related to an SCI theme or expertise. The regions covered by the SCIs include New England, Mid-Atlantic, South, Midwest, and the West.

The national SENCER project STEM education reform activities include the SENCER Summer Institutes, model courses, backgrounders, and a searchable digital library of course materials. The Summer Institutes are intensive workshops where interdisciplinary institutional teams work on SENCER-based curriculum development projects. Over the years, SENCER has identified over 30 model courses that exemplify the SENCER approach and provide faculty with course elements that can be broadly adapted and implemented (models are described in more detail below). To assist faculty in designing new courses, SENCER has produced an in-depth manuscript series called Backgrounders on topics ranging from nanotechnology to obesity. Finally, since the SENCER community has grown to be over a thousand, many faculty have developed SENCER course materials that are available through the SENCER website in the SENCER Digital Library.

SENCER MODELS

The SENCER project has identified over thirty SENCER model courses including Chemistry and the Environment described here. Through issues of civic consequence, a SENCER model course teaches science that is both challenging and rigorous. SENCER models typically engage students in the following ways:

- require students to engage in serious scientific reasoning, inquiry, observation, and measurement;
- connect scientific knowledge to public decision making, policy development, and the effective "work" of citizenship;
- respond to student interests including personal interests as well as public or civic ones;
- require students to engage in research, to produce knowledge, to develop answers, and to appreciate the uncertainty and provisionality of knowledge;
- intentionally work to achieve learning outcomes, such as quantitative reasoning and critical thinking;
- reveal high potential for significant enrollment and broad applicability, with room for local adaptation; and
- may connect to other institutional goals, such as increased civic engagement, improved personal choices, and behavior.

Many of the SENCER models are interdisciplinary with topics including HIV/AIDS, slow food, and sleep science. Numerous chemistry-related models have been identified and are listed in Table 2.1.

A SENCER MODEL: CHEMISTRY AND THE ENVIRONMENT

Chemistry and the Environment is designed to intentionally motivate and engage students by connecting basic chemical principles and civic engagement through public issues such as air quality, ozone depletion,

TABLE 2.1   Examples of Chemistry-Related SENCER Model Courses

Model	Author	Institution
Energy and the Environment	Trace Jordan	New York University
Global Warming	Sharon Anthony and Sonja Wiedenhaupt	Evergreen State College
Environment and Disease	Michael Tibbitts	Bard College
Brownfield Action	Peter Bower	Barnard College and Columbia University
Chemistry and Policy	Christopher Smart, Pinar Batur, and Stuart Belli	Vassar College
Coal in the Heart of Appalachian Life	Andreas Baur, Judy Byers, Galen Hansen, Erica Harvey, Debra Hemler, Phillip J. Mason, and Noel Tenney	Fairmont State University
Forensic Investigations: Seeking Justice through Science	Gregory Miller	Southern Oregon University
Renewable Environments: Transforming Urban Neighborhoods	Steven Bachofer and Phylis Cancilla Martinelli	Saint Mary's College of California
Chemistry of Daily Life: Malnutrition and Diabetes	Matthew Fisher	Saint Vincent College
Nanotechnology: Content and Context	Christopher Kelty and Kristen Kulinowski	Rice University
The Power of Water	Alix D. Dowling Fink and Michelle L. Parry	Longwood University
Uranium and American Indians	Catherine Middlecamp and Omie Baldwin	University of Wisconsin–Madison

global warming, acid rain, and water quality. Since the course allows students to fulfill the university natural science core requirement, the overall course learning objectives have been defined by university natural science core curriculum objectives. Specifically, students should acquire an ability to analyze scientific problems, generate logical hypotheses, evaluate evidence, and tolerate ambiguity.

The basic course content is typical for an introductory environmental chemistry course (Table 2.2). The syllabus is flexible and is modified to support the research projects. For example, when projects related to recycling are pursued, then topics related to polymers, plastics and paper chemistry are covered in detail. However, when biodiesel was the focus of a laboratory-based project (described below), then less time was spent on recycling-related topics and various organic chemistry topics were added. The syllabus is intentionally flexible and

TABLE 2.2   Typical Chemistry and the Environment List of Topics

Week—Quarter	Chemistry Topics	Civic Connection
Weeks 1 and 2	Chemistry basics	—
	Metric system	—
	Temperature units/conversions	—
	Periodic table	—
	Basic atomic structure	—
	Quantum mechanical model	—
	Atomic number and isotopes	—
	Atoms, molecules, and moles	Local water quality reports
	Unit conversions	—
	Chemical equations	—
	Stoichiometric calculations	—
Week 3	Atmospheric chemistry	—
	Cycles	—
	Composition	—
	Acids, bases, and acid rain	—
	Acids and bases	—
	pH scale	—
	Acid rain	Pollution allowances
	Coal burning and smelting	SO_2 Trading
Week 4	Ozone depletion	—
	Types of electromagnetic radiation	—
	Ozone layer	—
	CFCs	—
	CFC replacements	Montreal Protocol
Week 5	Photochemical smog	Clean Air Act
	Smog "road map"	Local air quality control boards
	Brown cloud and tropospheric ozone	Emission testing
Week 6	Global warming	Kyoto Agreement
	Greenhouse effect	Cap and trade
	Greenhouse gases	Carbon trading
	Absorption of IR radiation	—
	Feedback mechanisms	—
	Climate predictions	—
Week 7	Energy: fossil fuels, solar energy	—
	Basic of electrical production	—
	Coal: acid rain, smog, global warming	—
	Gasoline: smog and global warming	—
	Photovoltaic cells	—
	Passive and active solar designs	—

TABLE 2.2 *Continued*

Week—Quarter	Chemistry Topics	Civic Connection
Week 8	Nuclear energy	Federal nuclear regulations
	Types of nuclear decay	International regulations
	Radiation, roentgen absorbed doses (RADs), and roentgen-equivalent man (REM)	—
	Fission and fusion	—
	Chernobyl	—
Week 9	Plastics and paper	Recycling programs
	Polymers: types of plastics	—
	Plastics recycling	—
	Paper chemistry	—
	Paper recycling	—
Week 10	Hazardous waste	Waste management programs
	Household hazardous waste	—
	Low- and high-level radioactive waste	Medical waste processing
Biodiesel topics	Basic organic chemistry	—
	Hydrocarbons	—
	Functional groups	—
	Fatty acids and triglycerides	—
Water quality topics	Agricultural chemicals	—
	Fertilizers	—
	Pesticides	—
	Toxicology	—
	Water quality	EPA and Food and Drug Administration (FDA) regulations
	Water purification	Clean Water Act
	Reclaimed water	—

finalized after project topics are chosen. Students are always informed of changes and topics covered on exams are clear.

CAMPUS- AND COMMUNITY-BASED RESEARCH PROJECTS

The research projects have contributed our campus environmental resource assessment (similar projects are now managed by our campus Office of Sustainability) (Clugston and Calder 1999). The students choose a research project that is of interest to them and that connects to their daily lives. Project teams are typically four to six students and are set the first day of class. Since science majors occasionally enroll in the course, they are distributed among the groups. The groups are not typically assigned so students are allowed to form their own assemblies.[4] They are then provided time during class to meet once a week (at least 15 minutes). Each group prepares a project proposal during the first 2 weeks of the class. The project proposal is a collective effort and is a one- to two-page description of the project including specific objectives and relationship to previous campus-based studies if appropriate.

Within the first 4 weeks of the quarter, an individual preliminary report is required that describes specifically what each group member intends to do, any preliminary findings, and a timetable for the project as a whole and individual activities. A record of all individual project activities is maintained in a course journal or research notebook. In week 8, an individual progress report is due, which serves as a project update including background chemistry and relevant public policy information, description of methods if appropriate, and initial findings. The individual reports throughout the quarter provide a mechanism for monitoring each individual contribution to the project. At the end of the quarter (10 weeks), the final group report is submitted including the necessary background information, university-specific data and information, and recommendations for campus or community action with justification based on the research results. Each group also prepares a poster that is presented at a class poster session on the last week of classes. In most recent years, students have a choice of developing a web page or a poster. If a web page is developed, students present the page at the poster session. During the poster session, students stand by their poster in shifts so that all the students can view the posters as well as answer questions. In some years, administrators and others have been invited to the session. When other sustainability-related symposia are occurring on campus, students are also asked to present their poster at that event.

The specific learning objectives for the research projects are that students will[5]

1. develop an understanding of and appreciation for the practice of science;
2. learn basic science related to the project;
3. develop research skills:
   a. identifying a problem;

---

[4]Groups were assigned in the past, but due to the complexity of student schedules, it is easier to have students form their own teams.
[5]Op. cit. Shachter and Edgerly (1999).

    b. literature searching (previous campus work, literature or other sources, observations);

    c. developing hypotheses;

    d. proposing, designing, and conducting experiments;

    e. analyzing data; and

    f. interpreting results and drawing conclusions (campus recommendations);

4. communicate results in several formats (written, poster, Web, and oral);

5. gain time management and team management skills;

6. demonstrate an understanding of how the university operates;

7. develop sensitivity to the roles staff and administrators play in defining daily campus functions;

8. gain a sense of ownership and connectedness to the campus; and

9. recognize that they are environmental stakeholders in any place they choose to live.

The objectives related to research projects are similar to those typically linked to undergraduate research in science. Such experience has been demonstrated to successfully contribute to understanding the process of research (Russell et al. 2007), gains in communication and collaborative working skills, shifting from passive to active learning, as well as inspiring greater personal responsibility, independence, and ownership (Lopatto 2004; Seymour et al. 2004). Though the Chemistry and the Environment research projects are focused, intense 10-week experiences, many of the benefits attributed to undergraduate research experiences in the sciences are also observed for nonscience majors in this course. As demonstrated by narrative assessment tools,[6] Chemistry and the Environment students (74%) indicated that the projects helped them develop research skills and new methods of presenting and organizing results (65%). Offering research opportunities early in student careers may be a mechanism for attracting and retaining STEM majors (Russell et al. 2007) but, as shown in this course, also expands the understanding of citizens for the practice of science through experiences for nonscience majors.

As a form of civic engagement, the projects helped the students learn more about how the university operates (74%) and created a stronger feeling that the students have a greater stake in how the university

---

[6] Ibid.

operates (52%).[7,8] In a SENCER context, the Chemistry and the Environment content and structure allow students to learn chemistry fundamentals through the context of broader social questions focused on local applications. To inspire college students to be more civically engaged, "programs and organizations ought to address significant problems or passions in young people's lives, and preferably in the larger community in which they live. In addition, these efforts must allow young people to provide consequential input into decision-making and to produce tangible solutions or products ..." (Saguaro Seminar on Civic Engagement in America 2000). The primary Chemistry and the Environment structure for engagement is the chemistry-based research projects that are defined by the students and require students to make recommendations for addressing issues on campus or in the community and present those recommendations, supported by their research, to the community for consideration. Over the years, many, not all, projects have led to "tangible" results where students did have the opportunity to have "consequential input into decision-making" (Clugston and Calder 1999).

For the first eight years that Chemistry and the Environment was offered, no formal laboratory was associated with the course. When the course did not have a formal laboratory, projects occasionally involved laboratory experiments or various types of field testing. Two sample projects are described below, and a listing of addition project topics with chemistry and civic connections appears in Table 2.3 as well as in previous publications.[9,10]

Transportation A student-driven, chemistry-based research project focused on campus transportation issues including estimations of carbon dioxide and smog-related pollutants resulting from campus activities. The project related directly to the course discussions of global warming, smog, local air quality, and air pollution regulations framed by basic chemical concepts such as chemical composition of the atmosphere, basic chemical reactions (e.g., combustion processes and photochemical reactions), stoichiometry of

[7] Ibid.

[8] The SENCER project has refined the use of the Student Assessment of Their Learning Gains (SALG) tool for SENCER courses (http://www.salgsite.org/, accessed January 2009) and was only available in test formats when Chemistry and the Environment was most recently offered (see footnote 2 as well).

[9] Op. cit. Shachter and Edgerly (1999).

[10] Op. cit. Middlecamp et al. (2006).

TABLE 2.3   Research Projects *without* a Formal Laboratory Component

Project Focus	Chemistry Connections	Civic Connections	Recommendations
Potential hazards in photography laboratories including air quality and waste management	Atoms, molecule, ions Acids and bases Oxidation and reduction Chemistry of photography Toxicity of heavy metals and volatile organic compounds	Occupational Safety and Health Administration (OSHA) regulations Right to Know Act Federal, state, and local water pollution regulations	Better training for students, improving ventilation in photo laboratories
Office hazards including new carpets, copiers, and dry erase markers	Organic compounds Toxicity of volatile organic compounds and particulates Air quality testing	OSHA regulations Right to Know Act Indoor air quality monitoring and regulation	Use carpet materials that minimize outgassing and improve ventilation in campus copy center (no longer in operation)
Emissions and ground maintenance including lawn mowers, leaf blowers, and small vehicles	Energy and fuels Combustion reactions Atmospheric chemistry Acid rain, smog, and global warming	Federal, state, and local air quality standards and regulations Local air quality board Right to Know Act OSHA regulations	Transition to more electric vehicles; use electric powered blowers or other devices when possible; convert lawn mowers to propane or other cleaner burning fuels; use traditional tools such as push mowers, rakes, and brooms
Use of plastic bags by the campus bookstore	Organic compounds Polymers Plastics Recycling Biodegradable plastics	Waste reduction regulations	Place a recycling bin in the bookstore for used bags; offer rebate for students who bring their own bag; offer canvas bag options
Reducing electricity use on campus	Energy and fuels Electricity production Combustion reactions Atmospheric chemistry Global warming	Air quality regulations	Turn off lights; install motion sensors in residence halls; install energy-efficient lighting; place on-campus residents on an energy allowance

reactions (e.g., How much carbon dioxide pollution is produced [ideally] by a tank of gasoline? What is the role of fuel additives?), and driving forces of reactions (e.g., conditions of smog generation). An examination of smog and global warming also opened an avenue to further consider the health and economic effects in a chemical context (e.g., ozone as an "oxidizer" of lungs, leaves, and paint). Furthermore, discussion of the Clean Air Act opened the door to illustrating effective environmental regulation and also demonstrated how industries can adapt to environmental regulation such as the development of a catalytic converter (and the wealth of science in discussing how they work) and the removal of lead from gasoline. Finally, an essential component of the project was a set of recommendations for the campus related to transportation and supported by the research results that included incentivizing alternate transportation and carpooling. In-class discussions also focused on actions individuals could take to reduce air pollution (e.g., purchase a high-mileage vehicle [no SUVs]; use public transportation; reduce fireplace usage; and purchase a propane grill.

Indoor Air Quality A comprehensive investigation delved into several aspects of indoor air quality including radon in building basements on campus and volatile organic compounds in newly renovated offices. Some members of the team were from regions where radon has been a significant problem, and others worked in offices where "smells" from new materials had been a concern. The team contacted the campus environmental health and safety officer (EHSO) for assistance in identifying areas to test, gaining access to testing, assistance with testing, and resources for air quality testing. The project directly related to course discussions of atomic structure and nuclear chemistry (e.g., basic atomic structure, isotopes, radioactive decay, and ionizing and nonionizing radiation) as well as basic organic chemistry and toxicology (e.g., volatile organic compounds, paints, adhesives, exposure pathways). Since office areas were investigated, the students were able to explore occupational health and safety regulations and to gain an understanding of Right to Know laws. The EHSO was particularly helpful in interpreting federal, state, and local codes and regulations. One of the primary recommendations emerging from the project was that the campus should use only carpets, paints, and other new building materials that do not outgas volatile organic compounds. This work contributed to developing the current university practice to use only green building materials.

Other projects have focused on reclaimed water, desalination, alternative fuels for campus vehicles, environmental impact of campus development, recycling education programs, green materials in new residence halls, bottled water versus tap water, hazardous materials in the arts as well as in the sciences, use of recycled paper, pesticide use on campus, water conservation, solar energy options for the campus, paper towels and hot air dryers, composting of food waste, use of dehydration systems to reduce food service waste, and green materials for food containers. A broad range of possible projects provide students with the opportunity to become engaged campus and community citizens.

SAMPLE RESEARCH PROJECTS WITH FORMAL LABORATORY

In 2000, a formal laboratory was added to the course. The addition of a formal laboratory offered a tremendous opportunity to introduce various types of chemical analysis into the research projects. The laboratory also offered some challenges in that project experiments had to be designed by nonscience students with the guidance of the faculty instructor and the undergraduate laboratory assistant. To keep the research projects manageable for the faculty instructor (who taught both the lecture and laboratory components of the course), two strategies have been used. One approach was to narrow the project focus to some type of water quality issue with the students choosing topics of interest to them within that framework. The second approach was to have the students focus on one topic in the laboratory (biodiesel synthesis) with the students exploring various approaches and the class making an overall recommendation.

Water Quality Approach

Water quality projects such as those described below have been shown to be effective methods for engaging students in environmental chemistry courses for majors (Juhl et al. 1997) and for nonscience majors (Lunsford et al. 2007). When the water quality research projects were conducted, Chemistry and the Environment was linked to a world geography course as part of a learning community. Poor water quality and access to potable water were a global environmental theme for both courses. Consequently, the chemistry research projects focused primarily on water analysis. Field water testing kits, atomic absorption spectroscopy, and fluorescence methods (typically for biological contamination) were available for testing. To prepare students as they designed their experiments, the first few weeks of laboratory focused on understanding standards and calibration related to pH and temperature measurements as well as spectroscopic calibration curves. Students then used various field tests and spectroscopic methods. Results were compared to data available from governmental and nongovernmental organizations. Projects are described in Table 2.4 including the laboratory component associated with each project.

Furthermore, the connection to daily life is an important motivation for students as they define their research projects. Connections to the lives of students for some of the projects described in Table 2.3 are listed:

To Filter or Not to Filter? Many homes use water filters to purify tap water. These students lived either at home or in off-campus housing. Since the homes were older, the students were interested in the effectiveness of standard, readily available water filters and in the ability of filters to remove heavy metals associated with pipes in older homes. The students designed experiments to test water for lead and copper and evaluated how effective household water filters were in removing such metal ions. The students also investigated types of home filters and the methods of filtration.

Water Quality Assessment: Northern California Coast The students in this project were native Hawaiians and were interested in surfing. They were motivated to better understand and to explore in detail coast water quality. They designed and conducted a study to investigate biological and heavy metal contamination in coastal waters.

What Contributes to the Taste of Water? Bottled water has become quite common and many claim that bottled water tastes better. These students were interested in exploring if additives may contribute to a better taste in bottled water. The focus of this project was to begin to understand the "taste of water" in terms of the chemical constituents added to or commonly found in water. Iron and magnesium concentrations were studied.

Biodiesel Approach

A few months before the Chemistry and the Environment course was scheduled to begin, a student suggested exploring the use of biodiesel for campus trucks. The student worked for campus facilities and had received approval to convert one of the trucks to biodiesel. For

TABLE 2.4   Research Projects *with* a Formal Laboratory Component

Project Title	Chemistry Connections	Laboratory Component	Civic Connections
To Filter or Not To Filter	Atoms, molecules, and ions Essential metal ions Metal ion toxicity Water purification Home water filters both ion exchange and charcoal Water quality reports and EPA standards	Testing filtered and unfiltered tap water for lead and copper Testing new filters and used filters Atomic absorption (AA) methods for determining metal ion concentrations	Water quality including EPA regulations Local water quality reporting Home options for improving water quality
Water Quality Assessment: Northern California Coast	Ocean water composition Coastal pollution Sewage treatment Biological contamination Metal ion concentrations on coastal waters	Samples from six coastal locations Fecal coliform field testing Testing coastal samples for lead and copper ions AA methods for determining metal ion concentrations	Coast water quality regulations Surfrider Foundation and various county environmental health agencies EPA regulations
Semiconductors: Effects on Campus Water Supplies	Atoms, molecules, and ions Metals Conductors, semiconductors, and insulators Manufacturing semiconductors Groundwater contamination in Silicon Valley	Campus drinking water samples Aluminum, cadmium, and lead testing AA methods for determining metal ion concentrations	Superfund sites in Silicon Valley EPA regulations Electronic waste
What Contributes to the Taste of Water?	Atoms, molecules, and ions Ions in water Tap and bottled water sources and regulation Ions in water and taste	Campus drinking water samples Iron and magnesium ion testing AA methods for determining metal ion concentrations	EPA regulation of drinking water FDA regulation of bottled water

Chemistry and the Environment, the goal was to assist in determining an optimum method of making biodiesel and to recommend if the campus should make its own biodiesel. To achieve this goal, the Chemistry and the Environment laboratory was treated as a large research group with students working in laboratory pairs on different synthetic approaches. The conversion of cooking oil to biodiesel is a simple transesterification process and has been recently introduced into chemistry laboratory programs (Clarke et al. 2006; Bucholtz 2007; Stout 2007). Laboratory project grades were determined by a group proposal (hypothesis and experiments, two to five pages), individual laboratory notebooks, a brief written summary (two to five pages) by the group, and the presentation of results as part of an end-of-quarter research group meeting.

The laboratory began with students proposing a synthetic approach after reading a description of the biodiesel synthesis (Tickell 2000) and a class discussion. Students proposed experiments using various types of raw cooking oil (peanut, olive, corn, etc.) and used oil (filtered or unfiltered). Either potassium hydroxide or sodium hydroxide was proposed as the base, and methanol or ethanol was chosen as the alcohol. Each student pair had a unique combination of reactants. Students conducted the synthesis and prepared a report and presentation for the final "group meeting" at the end of the quarter. As part of the ending group meeting, the entire class identified optimal methods (new olive oil, potassium hydroxide, and methanol were the best combination) and the resulting biodiesel was poured into the converted campus truck for use that day. The group also prepared an overall recommendation: the process of making biodiesel would be too difficult on a large scale since filtering the used cooking oil was very difficult and large quantities of hazardous materials (strong bases and flammable solvents) were needed. The class recommended contracting with a local biodiesel facility to remove the used cooking oil and then to provide the campus with biodiesel. Since the campus has almost eliminated the use of diesel fuel, the recommendation was not implemented. However, the campus is trying to find a substitute for the use of diesel in emergency generators that are regularly tested. Biodiesel has not proven to be a viable alternative for such use since the generators are not used frequently and generator parts may be corrupted by the biodiesel. An investigation of alternative fuels for generators is an interesting future research project.

STRATEGIES FOR MANAGING RESEARCH PROJECTS

Research projects require careful management by the instructor. Over the years, a few strategies have been developed to more efficiently manage time and resources. First, starting the projects as early as possible and carefully refining project proposals is essential. Project outlines should be completed within the first 2 weeks, if not earlier. In reviewing the proposals, the instructor needs to work closely with the research team to craft a project that can be done within the quarter and with available resources. Students will often adopt a project of thesis scale and need guidance in focusing the effort to be productive within the quarter. In addition, students need assistance in understanding the equipment and resources available and need direction so that the same issues are not simply recycled year after year but are continually contributing to campus and community improvement. The instructor needs to meet with each research team at least once to discuss the project proposal and to work through project details.

A second strategy is to use undergraduate assistants to help manage the research projects. Students who have completed the course or interested science majors can work closely with research teams in designing and conducting research, especially laboratory components. For institutions with graduate students, teaching assistants can also be used.

To support the research projects and to reduce time stress on the instructor, the project topics must be integrated into the overall course content and structure. Intentionally linking course content and project topics allows the instructor to focus on one course rather than on one lecture course and a set of mini courses for each research team. To accomplish such synergy, the instructor must keep some aspects of the syllabus flexible to allow for project-related topics to be incorporated during the course. If the research topics are focused, such as on water or air quality, then related topics can be built into the syllabus initially. In addition, lecture time should be allocated to project management. Each week, at least 15 minutes of lecture time should be given to research group meetings. Such time allows the instructor to move around and talk with each group. The in-class meetings allow for a more efficient management of the projects for both students and faculty.

Finally, and perhaps most importantly, the instructor should establish partnerships with various campus constituencies to facilitate project execution. Instructors should develop collaborative relationships with the environmental health and safety offices, campus sustainability programs, housing offices, food service providers, and university operations including planning, facilities, maintenance, and purchasing. Over the years, such partners have been instrumental in defining and executing projects. For example, our EHSO has been instrumental in shaping and providing resources for research projects focused on indoor air quality and the use of hazardous materials. Our purchasing office has been a key source of information for paper purchasing practices and campus resource data. The instructor needs to establish partnerships to develop a support infrastructure for the projects so that the responsibilities can be distributed more efficiently.

OTHER FORMS OF ENGAGEMENT

In addition to research projects, various other course elements such as case studies and journals have been used to form intentional links to civic engagement. With the biodiesel laboratory project described above, case studies were also conducted as part of the lecture component of the course. A case study approach is a common tool in science teaching (National Center for Case Study Teaching in Science 0000).[11] In Chemistry and the Environment, students were asked to choose a local environmental issue for the case study where local could mean the university and surrounding community, a current hometown, or perhaps the town or city where the student was born. To assist in identifying an issue, students were asked to use two resources that link environmental issues to zip codes: the Green Media Toolshed website: http://www.scorecard.org/ (accessed January 2009) and the Environmental Protection Agency (EPA) website: http://www.epa.gov/epahome/commsearch.htm (accessed January 2009). These websites give students possible case study topics directly linked to a location of interest to the students—either where they currently live or where they grew up. Case studies from the course include

New York, NY: World Trade Center Disaster and Asbestos Contamination

Denver, CO: Ground Water Contaminants near Vasquez Boulevard and I-70

Sonoma, CA: Vineyard Development and the Environment

Alameda County, CA: Particulate Matter and Air Quality

Livermore, CA: Groundwater Contamination at the Lawrence Livermore Lab

Tempe, AZ: Indian Bend Wash Area

Seattle, WA: Environmental Impact of a Third Runway

Marin County, CA: Abandoned Gambonini Mine

Multnomah County, Oregon: Portland Harbor Superfund Site

San Francisco, CA: SFO Runway Reconstruction

Hayward, CA: Hayward Regional Shoreline

[11]An example: Dunnivant, F., A. Morre, M. Alfano, R. Brzenk, P. Bickley, and M. Newman (2000) Understanding the greenhouse effect: Is global warming real? An integrated lab-lecture case study for non-science majors. *J. Chem. Educ.* 77, 1602–1603.

Chicago, IL: Chicago River Case Study

San Francisco, CA: Mercury Contamination of San Francisco Bay

Mojave, CA: Edwards Air Force Base and Hexavalent Chromium

St. Louis, MO: Times Beach and Dioxin

San Jose, CA: Mercury in the Guadalupe River

Case studies involved a proposal (1–2 pages) and a final written report (10–15 pages). Each final report consisted of the following sections: (1) nature and origin of the problem; (2) history and current status (including contacting local officials or agencies); (3) exposure pathways, biological effects on living things, and health risks (Agency for Toxic Substances and Disease Registry, http://www.atsdr.cdc.gov/); (4) standards and regulations (http://www.epa.org/ or state EPAs such as http://www.calepa.ca.gov/); and (5) management, remediation, or other response. As an example, we assembled a case study in class on perchlorate contamination in local wells.

In addition to case studies, a course journal has also been an effective tool. Students were asked to complete problems related to basic chemistry topics and internet resources in the journal. For example, end-of-chapter problems in the textbook[12] were often assigned and completed in the journal. Students were also asked to record comments and questions related to course materials and to maintain a project log. Specifically, journal entries included responses to and analysis of internet resources, responses to and analysis of lecture material, critical responses to periodical articles and government documents, project ideas, and project progress including data, interviews, and event logs. The journals were collected at least one time during the quarter and at the end of the quarter.

CONCLUSION

Chemistry and the Environment is a SENCER model course for non-science majors designed to teach basic chemistry through projects, case studies, and other course elements linked to topics of civic interest.

---

[12]Textbooks have included Buell, P. and J. Girard (2002) *Chemistry: An Environmental Perspective*, 2nd ed. Boston: Jones and Bartlett Publishers and Baird, C. (1999) *Environmental Chemistry*, 2nd ed. New York: W.H. Freeman and Company (new edition in 2008 with coauthor M. Cann).

The research projects are themselves a form of civic engagement and contribute to campus and community sustainability. The course is a SENCER model that offers paths for students to embrace civic engagement and responsibility while learning science.

REFERENCES

Bucholtz, E. (2007) Biodiesel synthesis and evaluation: An organic chemistry experiment. *J. Chem. Educ. 84*, 296–298.

Clarke, N., J. Casey, E. Brown, E. Oneyma, and K. Donaghy (2006) Preparation and viscosity of biodiesel from new and used vegetable oil. *J. Chem. Educ. 83*, 257–259.

Clugston, R. and W. Calder (1999) Critical dimensions of sustainability in higher education. In *Environmental Education, Communication and Sustainability*, Vol. 5, ed. W. Filho. Frankfurt: Peter Lang.

Juhl, L., K. Yearsley, and A. Silva (1997) Interdisciplinary project-based learning through an environmental water quality study. *J. Chem. Educ. 74*, 1431–1433.

Lopatto, D. (2004) Survey of Undergraduate Research Experiences (SURE): First findings. *Cell Biol. Educ. 3*, 270–277.

Lunsford, S., N. Speelman, and A. Yeary (2007) Characterizing water quality in students' own community. *J. Chem. Educ. 84*, 1027–1030.

National Center for Case Study Teaching in Science (2009) http://library. buffalo.edu/libraries/projects/cases/case.html (accessed January 2009).

Russell, S., M. Hancock, and J. McCullough (2007) Benefits of undergraduate research experiences. *Science 316*, 548–549.

Saguaro Seminar on Civic Engagement in America (2000) Better Together. Report of the Saguaro Seminar on Civic Engagement in America, p. 80, John F. Kennedy School of Government, Harvard University.

Seymour, E., A.-B. Hunter, S. Laursen, and T. DeAntoni (2004) Establishing the benefits of research experiences for undergraduates: First findings from a three-year study. *Science Education 88*, 493–534.

Stout, R. (2007) Biodiesel from used oil. *J. Chem. Educ. 84*, 1765.

Tickell, J. (2000) *From the Fryer to the Fuel Tank: The Complete Guide to Using Vegetable Oil as an Alternative Fuel*, 3rd ed. Covington, LA: Tickell Energy Consultants.

CHAPTER 3

CORD's *Applications in Biology/ Chemistry*: Teaching Science in the Context of Major Life Issues

BONNIE RINARD and MARK WHITNEY

Center for Occupational Research and Development (CORD), Waco, TX

We believe, after examining the findings of cognitive science, that the most effective way of learning skills is "in context," placing learning objectives within a real environment rather than insisting that students first learn in the abstract what they will be expected to apply. (From the Executive Summary of *What Work Requires of Schools* from the Secretary's Commission on Achieving Necessary Skills, U.S. Department of Labor, June 1991)

In the center of a city of 400,000 people is a large park, with a huge, spring-fed public pool.[1] The pool is bordered by the natural rock formations of the creek that feeds it, but with a concrete wall along one end. Big trees shade the hills along the half-mile stretch of springs. The water is very cool, the city's best offering during the dog days of summer. But today, the creek is quiet. A few miles away, the park's usual patrons

---

[1] The scenario that runs through this chapter is based on a multiple-part scenario found in the *Water* unit of CORD's *Applications in Biology/Chemistry* (ABC) series. The other titles in the series are listed at the conclusion of the chapter. The authors wish to express their indebtedness to two prior CORD publications on contextual teaching and learning: *Teaching Contextually: Research, Rationale, and Techniques for Improving Student Motivation and Achievement in Mathematics and Science* (Crawford 2001) and *Promising Practices for Contextual Learning* (Harwell and Blank 2001).

Making Chemistry Relevant: Strategies for Including All Students in A Learner-Sensitive Classroom Environment, Edited by Sharmistha Basu-Dutt

have gathered outside the city hall to protest the planned development of land that is part of the creek's watershed. If this area is developed as planned—with a high-tech industry, a golf course, and a housing development—environmentalists predict that the creek will receive so much polluted runoff it will soon be unfit for recreation. The citizens are pressuring the city council to refuse the developers a permit to further develop the watershed. Among those at the hearing are a high school biology teacher, Joanne Li, and two of her students, Leroy Davis and Veronica García.

Whether they realize it or not, Leroy and Veronica have just embarked on an unforgettable process in which they will learn valuable life lessons through hands-on experience, lessons pertaining to one of the most basic and necessary substances in the human experience— water. They will learn about the unique properties of water, how water is used, the chemistry of water, and how water use regulations affect not only the environment but also virtually every aspect of every person's life, in the home, in the workplace, and in society in general. In short, Leroy and Veronica are about to experience *contextual teaching and learning* (CTL) at its best.

This chapter will describe certain aspects of CTL and will show how the units that make up CORD's ABC series create learning environments in which CTL can fulfill its potential. We will focus particularly on the *Water* unit (ISBN 1-555-02-837-3), from which the narrative, or "scenario," involving Leroy and Veronica was taken. (Every unit in the series has a strong narrative element that enables students to see how the concepts under consideration are used outside the classroom.)

OVERVIEW OF CTL

As its name makes clear, CTL is all about *context*. CTL can take many forms but always involves a process in which students encounter new information in contexts that are already familiar to them or in which they can readily see how the information is useful. CTL enjoys the support of research in cognitive science (as summarized in Crawford [2001] and in Harwell and Blank [2001]). Moreover, the experiences of many educators have shown that CTL produces better results than the drill-oriented, stimulus-and-response methodologies that have dominated American education for many years.

CTL is effective because it is consistent with human nature. People internalize new information most readily when that information makes sense to them and is perceived as valuable. CTL helps students to discover the relationships that the human brain naturally seeks (Jensen 1995). CTL is also effective because it is based on recognition of the fact that, ultimately, all learning is *connected* and that the natural connections sought in the teaching and learning process extend beyond the classroom. Learning can take place in many kinds of settings—classrooms, laboratories, workplaces, social events, family settings, and civic occasions, to name a few. In each of these settings, daily life obstacles and challenges present themselves. CTL encourages, organizes, and rewards the natural human impulse to solve those problems.

Research has shown, and every teacher has observed, that human intelligence varies not only in degree but also in kind. There are numerous *types* of intelligence. Accordingly, people learn differently—some by absorbing facts or abstract concepts, others through hands-on activities in which facts and concepts are demonstrated or "worked out." CTL accommodates, even celebrates, the natural diversity of learning styles in the classroom. This is true with respect not only to teaching but also to assessment. Because people learn differently, the most effective means of assessment is one in which different learning styles are accommodated. CTL naturally lends itself to that accommodation. In an era in which heavy emphasis (some would say too heavy) is placed on fixed-response standardized tests, CTL provides "poor test takers" a better opportunity to demonstrate acquired knowledge and skills.

In their exploration of "brain-based learning," Renate and Geoffrey Caine established 12 principles that provide a cognitive base for understanding why students are successful in CTL environments (Harwell and Blank 2001).

Principle 1: *The brain is a complex, dynamic system.* CTL seeks an appropriately broad base of teaching activities.

Principle 2: *The brain is social.* CTL encourages cooperation among learners.

Principle 3: *The search for meaning is innate.* CTL helps learners to find meaning through application.

Principle 4: *The search for meaning occurs through "patterning."* CTL encourages students to discover patterns that enable the transfer of knowledge from familiar to unfamiliar contexts.

Principle 5: *Emotions are critical to patterning.* By focusing on personal relevance, CTL encourages students to focus on issues in which they are emotionally "invested," thereby enriching the learning experience.

Principle 6: *Every brain simultaneously perceives and creates parts and wholes.* CTL encourages students to see how each newly acquired fact relates to broader contexts.

Principle 7: *Learning involves both focused attention and peripheral perception.* CTL naturally seeks situations in which students can interact with teachers and other professionals who are passionate about their work, thus providing "peripheral" stimuli that encourage learning.

Principle 8: *Learning always involves conscious and unconscious processes.* CTL improves student performance by encouraging active processing of information.

Principle 9: *We have at least two ways of organizing memory—spatial memory and rote learning.* When information is merely memorized, without being rooted in experience, the information is soon forgotten. CTL encourages students to seek connections between information and experience.

Principle 10: *Learning is developmental.* CTL helps students to see logical progressions from the familiar to the unfamiliar, from simple to complex, and from concrete to abstract (see also Harwell 1999).

Principle 11: *Complex learning is enhanced by challenge and is inhibited by threat.* CTL is conducive to the creation of learning environments that are challenging but are nonthreatening.

Principle 12: *Every brain is uniquely organized.* CTL accommodates the natural diversity of learning styles.

In recent times, more and more educators (and educational researchers) have come to the realization that people naturally "construct" knowledge based on what they already know and believe (Bransford et al. 1999). CTL supports the "constructivist" classroom by encouraging students to construct their own knowledge by solving problems and by explaining their solutions rather than by merely memorizing and reciting facts. CTL is not "dumbed down" teaching, as some of its critics contend. On the contrary, CTL provides a framework for establishing educational settings in which students can meet high academic standards. It "works" because it helps students to learn in the way that all people learn best—by making connections between their lives and the knowledge and information they are exploring. Those connections enable students to build a solid foundation that serves them well beyond the classroom and well beyond their years of formal schooling.

CORD'S PHILOSOPHY: THE "REACT" METHODOLOGY

At the outset of this chapter, we met two high school students, Veronica and Leroy, whose experience with CTL is told in the ABC *Water* unit.

Veronica and Leroy have become involved in a public debate over the proposed development of Hondo Creek, a popular recreational spot in their community. As their story unfolds, they make the astute observation that the city officials who will ultimately decide the fate of Hondo Springs are not especially well informed about the science involved in water conservation and use. Clearly, the council members need more information. So, Veronica and Leroy, under the guidance of their teacher Joanne Li, decide to educate the council during the next round of hearings. They begin by doing library research on water. They read about water in the leaves of plants, water in cells, and water in kidney function. Then they move their research outdoors and focus on the springs. They conduct a survey of life forms in and around the springs—plants, fish, land animals, and microscopic life.

As Veronica and Leroy explore the chemistry of water—and as students work their way through the *Water* unit—they engage in a process that involves five distinct but interrelated elements:

- **R**elating (learning in the context of life experiences);
- **E**xperiencing (learning in the context of exploration, discovery, and invention);
- **A**pplying (learning in the context of how knowledge and information can be used);
- **C**ooperating (learning in the context of sharing, responding, and communicating with other learners); and
- **T**ransferring (using knowledge in new contexts or novel situations).

Collectively, these five elements make up what CORD, the developer of the ABC series, refers to as the REACT methodology, named after the acronym for the five elements encompassed. CORD's teaching materials are specifically designed to support constructivism in the classroom by facilitating the creation of learning environments in which these five elements are encouraged. When the REACT methodology is used effectively, students do not just remember, they *understand*.

Let us look at each of the five elements individually.

Relating

In constructivist learning environments, *relating* takes place when students see meaningful connections between "new" (i.e., unfamiliar) concepts and contexts that are familiar to them. They connect something they do not know yet to something they already know well. When this happens, the students experience what Caine and Caine call "felt meaning." The "aha" moment that accompanies felt meaning can make a lasting impression on students because it enables them to see useful-ness and meaning in concepts that had previously been confusing or pointless (Crawford 2001). Relating is natural, but it does not happen automatically. In the classroom, it must be planned and encouraged by the teacher. For example, every student has knowledge, memories, and associations that are relevant to the study of the chemistry of water, but few students will see the relevance of their own experiences without encouragement and guidance from the teacher.

Consider the scenario in which Leroy and Victoria have become involved. Before they can take a clear position in the water use con-troversy taking place in their town, they must give serious thought—probably for the first time—to how water is used outside their homes. In a more traditional, lecture-based classroom setting, their teacher might present a chart showing water use statistics for a variety of normal human activities—drinking, sanitation, manufacturing, recre-ation, and so on. In the CTL classroom, the teacher introduces a fact-finding process to which *all* students can relate, since everyone needs and uses water.

One of the most effective ways the ABC units encourage relating is by posing questions that encourage students to focus first on familiar concepts, then transitioning into unfamiliar territory. Every subunit of every ABC unit begins with thought-provoking questions involving familiar contexts. For example, when students turn to the beginning of the first subunit of the *Water* unit, they see four pictures: a young man taking a drink of water from a glass, a young woman in a lab coat examining a test tube containing water, a motorboat on a lake, and a hydroelectric dam. Beneath the pictures are the following questions and prompts:

- Describe the way water is used in each instance.
- Can you think of another liquid that would substitute for water in each instance? Why?
- What is unique about water?
- What makes it useful?

Almost any student could identify the uses of water depicted and would have a sense of why no other liquid would serve the same purpose. (For example, any student could name liquids that would be unpalatable, even dangerous, to drink.) But when asked "why" water is uniquely suited for the purposes depicted, many students would have no answer because they do not yet understand the *chemistry* that gives water its unique properties. The questions posed would naturally lead to others: "Where does our water come? What factors affect the quality of our water? How much influence does human activity have on the availability and quality of our water?" One goal of having students focus their attention on the uses of water is to get them to see how those uses may vary depending on what *phase* the water is in. This represents another area in which students will be familiar with certain phenomena—what happens to water when it is exposed to temperature extremes—but few students will be able to explain why those changes take place. So, again, we see that the progression from familiar to unfamiliar creates an environment in which every student can participate. Throughout the relating process, the teacher reinforces the students' prior knowledge by summarizing student answers, by elaborating on their examples, and by asking questions that help them to clarify and refine their thinking.

Experiencing

As described in the preceding section, relating connects new information to life experiences or prior knowledge that students bring with them to the classroom. The challenge of taking students to the next level of understanding involves *experiencing*, in which the students explore, discover, and invent. Experiencing—which we might also describe as *learning by doing*—typically involves hands-on activities and laboratories in which students use manipulatives to explore abstract concepts concretely.

Activities   Recall that Veronica and Leroy recognized the need to educate their city council members on water usage issues, a need that can only be met by finding relevant data. The process that Veronica and Leroy undertake represents a fundamental aspect of experiencing. They are not provided with "canned" data and are not asked to manipulate it. They must acquire their own data via research—a laborious but rewarding experience that brings the learning process to life. One of the arguments made in favor of developing Hondo Creek is that, without it, the town would become a "ghost town." Leroy and Veronica naturally wonder whether that argument is valid. Only hard facts and figures will tell.

The ABC units provide many opportunities for students to gather data on real-life issues. The following is a sample taken from the *Water* unit (Fig. 3.1). Notice that the activity does not ask students to gather information about water use only in familiar contexts (home, school,

---

Activity 1-3

- Conduct an informal poll of adults you know (parents, neighbors, friends). Find out how water is used in their workplaces. Ask them to be as specific as possible in their answers. Also, try to ask people in a variety of occupations.

- As a class, compile the results of your informal poll to come up with a list of ways water is used. Add to the list any uses you can think of that take place at school, in the home, or in recreational areas.

- In small groups, examine the class list of water uses and try to group the water uses in some way that makes them easy to understand. In other words, develop categories based on the way water is used.

- Compare the categories developed by your group with those created by other groups. As a class, decide which system you like best and why. Keep your list and categories in your ABC notebook for use in later activities.

Important note: Keep in mind that your class poll is not likely to represent a cross section of the entire community. To find out who the major users of water are in the community, you would have to ensure that your sample represented all sectors. Your poll is, however, a good starting place. (ABC *Water*, p. 8)

FIGURE 3.1   Sample activity from ABC *Water*.

recreation) but also in the workplace, a context that will probably introduce them to water uses they had not previously considered.

In other activities in the *Water* unit, students are asked to gather information about contexts in which water is used as a solution, about the industrial uses of water, and about the acid–base properties of foods. One series of activities asks the students to conduct water habitat surveys of bodies of water in their communities and to srecord information about the physical and chemical properties of those bodies and the life forms that inhabit them.

The ABC activities engage students' creativity in solving real-world problems. In the activity presented in Figure 3.2, students explore water use issues in the ranching business.

The ABC activities provide a meaningful way for students to learn new concepts. The activities are specifically designed to create in the students a sense that the topics under discussion represent concepts that they *need to know*. As Veronica and Leroy explore water uses in their community, they discover (with the help of their teacher) that they cannot complete their research without learning about acids and bases, new concepts (for them) to which they had never given serious thought. This discovery initiates a series of activities designed to help students understand the *usefulness* of information about acids and bases and other relevant concepts. In one activity, the students perform inventories of the acid content and characteristics of the foods in their

Activity 2-10

- Assume that you are going into the ranching business. You don't have a lot of experience, so you start off on a small scale. You buy a 150-acre ranch and 100 heads of cattle. The land has no stock pond or tank to provide the herd with drinking water. The land is mostly sand in some areas and mostly clay in others. You know that the land has a natural spring as well as a plentiful underground water supply you can tap. How can you best supply your herd with a year-round source of clean drinking water?

- Research your problem by talking to a local agricultural extension agent, an animal husbandry teacher, or a livestock veterinarian.

- In your ABC notebook, develop a plan or a design for your solution.

- Share your solutions with the class. Discuss the solutions, their advantages, and their disadvantages. As a class, select two solutions you think will work best and be cost-effective. (ABC *Water*, p. 59)

FIGURE 3.2   Sample activity from ABC *Water*.

homes. In another, they extend their surveys to include hydroxide content. In yet another, they explore ion exchange as a method of water purification.

Throughout the ABC units, the activities are reinforced with occupational scenarios. In the *Water* unit, students take a virtual tour of a fertilizer plant, where they learn about sulfuric and phosphoric acids and other facets of the chemistry involved in the manufacture of fertilizer. At the conclusion of the tour, the guide summarizes the importance of the experience in this way: "Next time you look at a bag of fertilizer and see 'phosphates' in the list of ingredients, you won't take it for granted." The ABC units also reinforce student activities with career profiles. In the *Water* unit, in a section pertaining to acids and bases, students meet Rosalinda, a chemist. Rosalinda works in her state's hazardous chemical emergency response unit, which houses the state's "hotline" for hazardous chemical cleanup. As Rosalinda explains, determining the appropriate first response to hazardous spills involves a knowledge of chemistry. Other career profiles in the unit introduce students to a power plant chemical technician, a hemodialysis technician, cotton farmers in West Texas, a phlebotomist, a pharmaceutical technician, a hydrographic survey technician, a hydrogeologist, and a wastewater treatment technician. Many of the career-centered features of the series contain a strong environmental focus. (Two of the units, *Natural Resources* and *Waste and Waste Management*, are devoted almost entirely to environmental issues.) In the *Water* unit, the entire final subunit focuses on protecting the quality of water.

Topics within the unit include sources of water pollution, natural processes that affect the quality of water, the effects of human activity on water quality and quantity, and water treatment technology.

In addition to giving real-world meaning to scientific concepts, the ABC activities teach problem-solving skills, analytical thinking, communication, and group interaction.

Laboratories At the end of almost every ABC subunit are one or more laboratories. While placed at the end of the subunit, the laboratories are to be done early in the sequence so students can explore and experience the concepts covered in the subunit. The laboratories are like the activities but are usually longer and require more planning and more complex manipulatives. In the laboratories, students work in small groups to collect data by taking measurements, analyzing data, drawing conclusions, making predictions, and reflecting on the fundamental concepts involved.

Table 3.1 presents sample laboratory topics from the *Water* unit, along with an indication of some of the manipulatives used and the real-world connections involved.

Applying REACT

The *applying* element of the REACT methodology involves *learning by putting concepts to use*. In the ABC series, the application exercises often take the form of "word problems," not unlike those found in many textbooks. But they are different in two fundamental respects: First, they pose realistic (often occupational) situations. Second, they demonstrate the usefulness of academic concepts outside the classroom. Both of these differences add motivation to the learning experience. Consider the following, a typical word problem from a chemistry lesson on pH.

> The hydronium ion concentration in a certain aqueous solution is determined to be 1×10^{-4} mole per liter. (a) What is the pH of the solution? (b) What is the hydroxide ion concentration of this solution?

The intent of the problem is to have students understand the definition of pH and to use the fact that, in water and dilute aqueous solutions and within the ordinary range of room temperatures, the ion product $[H_3O^+] \times [OH^-]$ of 1×10^{-14} is commonly used as a constant. If you know the concentration of either the hydronium ion or the hydroxide ion, you can calculate the concentration of the other. But apart

TABLE 3.1   ABC *Water* Laboratory Activities, Manipulatives, and Real-World Connections

Laboratory	Activities and Manipulatives	Real-World Connections
Movement of Molecules across a Membrane	Students use dialysis tubing to study osmosis between water and various concentrations of sodium chloride.	Water treatment
How Does Antifreeze Work? (Parts 1 and 2)	In Part 1, students investigate the effect of various concentrations of antifreeze on the freezing point of water. In Part 2, students look at the effect of antifreeze on the boiling point of water.	Automobile maintenance
How Is Density Measured?	Students use a hydrometer to measure the specific gravity of the various antifreeze solutions used in the previous laboratories.	Automobile maintenance
How Does pH Affect Hydroponic Plant Growth?	Students grow plants hydroponically in solutions of differing pH and observe the results.	Agriculture
How Is pH Lowered? How Is pH Raised?	Students use a pH meter to measure the effect of adding drops of 0.1 N hydrochloric acid or 0.1 N sodium hydroxide to distilled water.	Swimming pool maintenance
What Do Buffers Do?	Students titrate a weak acid with a strong base and observe the pH changes during the titration using a pH meter.	Acid rain technology
Testing Water Quality	Students use pool and spa test kits to test water for pH, chlorine, alkalinity, and hardness.	Swimming pool and spa maintenance

from some meaningful context, the problem implicitly invites the student to wonder, "So what?"

The ABC units answer the "so what" question by providing contexts that students can relate to. In the *Water* unit, for example, a laboratory entitled "How Is pH Lowered?" begins by introducing a fictional character who lowers pH as part of her job:

Anita K. is a lifeguard at a large public swimming pool. One of her jobs is to test pH and chlorine levels in the pool four times a day. She enters these measurements into a log. "Chlorine and pH levels in a swimming pool are interrelated," says Anita, "When the chlorine gas we add to the pool water dissolves, it forms the hypochlorous acid ($HClO$) and hydrochloric acid (HCl). It is primarily hydrochloric acid that lowers the pH of the water in the pool." The pool where Anita works also receives sodium hydroxide ($NaOH$) solution, which neutralizes hydrochloric acid.

Today, when Anita checked the pH and chlorine levels she found an acceptable chlorine level, but a pH of 7.8, which is *above* the acceptable level. She enters this value in the log and goes to the pump house to adjust the pumping rate on the sodium hydroxide. (ABC *Water*, p. 159)

Because of the real-life context that introduces the laboratory, students see the importance of the key concepts involved in adjusting the pH. To ensure that students see that this is but one of *many* contexts in which the ability to adjust pH is relevant, the *Water* subunit on acids and bases includes (under "Activities by Occupational Area") additional activities in which students are introduced to real-life applications in five broad categories:

- general (example activity: compare pH levels in shampoos),
- agriscience (example activity: interview local agricultural extension agent and area growers to find out about the alkalinity vs. the acidity of local soils),
- health occupations (example activity: explore the concept of blood pH via interviews with local hospital staff members),
- family and consumer science (example activity: explore the use of chemical salts in the manufacture of textiles), and
- industrial technology (example activity: obtain information from local print shops on how acid is used in printing and printmaking).

These are sample activities of one particular subunit of the *Water* unit. Every subunit in the ABC series provides comparable occupation-specific activities for the topics under discussion. The ABC application exercises and activities have proven effective because they are realistic and are applicable to a broad range of students' current and possible future lives outside the classroom (as consumers, family and community members, employees, and citizens).

The relating and experiencing components of the REACT methodology provide strategies for developing felt meaning, understanding, and insight—empowering mental phenomena that help students to see that they *can* learn the information being presented. The applying element helps students to develop a deeper sense of meaning—a reason for learning. It shows students not only that they *can* learn a given body of information but that they *should* learn it, as a means of enhancing their own lives. Collectively, the relating, experiencing, and applying elements are highly motivational.

Cooperating

Because they involve realistic situations, many of the problem-solving exercises in the ABC series are complex—more so than most students would be able to handle working alone. As we follow Veronica and Leroy in researching Hondo Springs, we notice that many of the insights they develop are the result of their brainstorming sessions together, sometimes under the guidance of their teacher, sometimes not. The developers of the ABC series consider this kind of interaction an essential component of life outside the classroom that should be reflected in the learning environment. Consequently, many of the activities and laboratories in the ABC series are designed to encourage the fourth element of the REACT methodology, *cooperating—learning in the context of sharing, responding, and communicating with other learners.*

For example, in one activity in the *Water* unit, students are asked to develop a storyboard for a soap commercial. In another, student groups discuss and research various aspects of agricultural uses of water. In the group environments encouraged by the ABC materials, students are less self-conscious and are confident that they can voice their opinions or ask questions without fear of embarrassment. They are also more willing to explain their understanding of key concepts or to recommend problem-solving approaches. Peer interaction encourages students to reevaluate and reformulate their own thoughts and to value the opinions of others. When student groups succeed in reaching common goals, their success becomes a shared experience.

In the area of cooperative learning, the ABC materials are consistent with research. Many studies show that cooperation among students, as opposed to more traditional individual- and competition-based methods, promotes higher student achievement (Pintrich and Schunk 1996). The ABC group activities are also consistent with established guidelines for conducting cooperative learning experiences. For example, they help to create environments in which each student feels that individual success is ultimately dependent on group success. They promote discovery and reasoning, the generation of new ideas and solutions, and the transfer of strategies and facts learned within the group to problems considered individually. They also provide opportunities for students to acquire interpersonal and small group skills such as leadership, decision making, trust building, communication, and conflict management.

The cooperative learning aspect of the ABC series places special demands on the teacher, who must be able to form well-balanced groups, assign suitable tasks, monitor group progress, diagnose problems within groups, and supply whatever information or direction is necessary to maintain group progress. Like the other elements of the REACT methodology, *cooperating* is challenging but worth the effort.

Transferring REAC**T**

In the conventional classroom, the teacher's main function is to present facts and to describe procedures. The students then memorize the facts and practice the procedures, usually through "skill drills" and (less often) word problems. In a constructivist (contextual) classroom, on the other hand, the teacher's function is broader and focuses on understanding rather than on memorization. Constructivist teachers use relating, experiencing, applying, and cooperating (the REACT elements discussed thus far) and assign experience-based, hands-on activities and realistic problems that enable students both to gain initial understanding and to deepen their understanding.

When the first four elements of the REACT methodology are effectively encouraged, students are able transfer what they learn to unfamiliar contexts. *Transferring* (the final REACT element) takes place when *knowledge is used in a new context or novel situation—one that has not been covered in class.*

The ABC materials encourage *transfer* in numerous instances. Particularly good examples are found in the "Challenge" sections that conclude the laboratories. The *Water* unit, for example, includes a laboratory designed to help students understand how molecules move across membranes. The stated purpose of the laboratory is to help students "determine the effect of concentration difference on the movement of water and solute across a membrane." The laboratory's stated objective is to enable students to "predict the direction of material movement across a membrane based on the concentration of materials on both sides of the membrane." During the laboratory, students measure mass with a balance and work with dialysis bags. At the conclusion of the laboratory, students explore questions designed to help them transfer what they have learned to contexts outside the classroom:

- In a reverse osmosis unit, how would the purity of the water in the outlet tank be affected if the high-pressure pump lost power?
- Based on your data, what would happen to the body fluids of a marine fish in a freshwater lake?
- Based on your data, what would happen to the body fluids of a freshwater fish that swam into the Atlantic Ocean?

Similarly, in another laboratory in the *Water* unit, students "determine the relative melting points of a series of antifreeze solutions of different concentrations." In carrying out the laboratory, they measure the density of a liquid, relate density or specific gravity to concentration, and make choices about antifreeze concentrations (and explain their choices). At the end of the laboratory, the students are asked to *transfer* what they have learned about antifreeze to a horticultural context, via a thought-provoking "challenge" question: "When the solution inside a cell freezes, the solution may expand, rupturing the cell membrane. What kind of plant-breeding program would develop plants that are less vulnerable to this kind of freeze damage?"

When the REACT methodology is effectively applied using the ABC materials, the result is creativity, strong student motivation and engagement, enjoyment, communication, and positive group interaction. Using these materials is not necessarily easier than using more conventional materials, since constructivist teaching may require additional prep time and effort, especially for teachers who are accustomed to using more conventional methods. The constructivist approach requires dedication and persistence on the part of the teacher. Becoming comfortable with constructivist approaches is a gradual process that may also require long-term professional development, mentoring, practice, collaboration, and reflection. However, the results justify the necessary expenditure in time and effort.

The day after a critical vote on Hondo Springs, Veronica and Leroy examine the local newspaper in class. The ordinance that was finally passed provided a reasonable amount of environmental protection without stifling economic development—a true-to-life compromise. Best of all, Veronica and Leroy have the satisfaction of knowing that they played a role in making sure the decision was based on accurate information. Their experiences of the last few weeks have made an impression that will last a lifetime and, who knows, may lead them to successful careers in natural resource management or some related field. That is CTL at its best.

CHALLENGES AND REWARDS

Using the REACT methodology is not easy, for teachers or students. The REACT methodology looks beyond memorization of facts and figures and challenges students to find meaning and to seek patterns and relationships. For many teachers (perhaps most), creating classroom environments in which the REACT methodology can be effectively implemented requires a new kind of preparation and a willingness to let go of the conventional lecture-style, chalk-and-board teaching that has been the norm for many years. The REACT methodology often involves manipulatives that must be on hand when students arrive in the classroom, and it often challenges teachers to acquire expertise that falls outside their areas of training—business, technology, health care, communication, manufacturing, agriculture, and many others. For example, for many teachers, even experienced high school chemistry teachers, leading students through a scenario such as the one described in this chapter would involve a learning process.

Any REACT-supportive classroom environment will be characterized by the following traits, all of which involve a degree of "outside-the-box" thinking for most teachers:

- Communication is not directed entirely by the teacher but is reciprocal.
- Concepts are presented in real-life contexts in which students can see the usefulness of the information they are being asked to learn.
- Concepts are presented in contexts that are related to what the students already know and consider important.
- Students spend more time engaged in their own work than listening to the teacher talk.
- Classroom activities and assessment methods accommodate multiple learning styles.
- Information is not learned first and applied later. It is applied as it is learned. The application is what makes the learning possible and long-lasting.
- The teacher's words usually take the form of questions rather than directives.
- Questions posed by the teacher do not have single-word answers.
- Students feel comfortable turning to each other for help.
- Students are encouraged to find answers to their own questions.

- The teacher is a facilitator rather than a dispenser of information.
- Students collect their own data rather than work with hypothetical data provided by the teacher or by textbooks.
- Students are encouraged to explore and to understand how the discipline in which they are immersed relates to other disciplines. To an extent, all REACT-based teaching is interdisciplinary.

Change in teaching practice is difficult to bring about. In 2000, the National Commission on Mathematics and Science Teaching for the 21st Century noted that

> Despite the dramatic transformations throughout our society over the last half-century, teaching methods in mathematics and science classes have remained virtually unchanged. Classroom practice has still hardly begun to capitalize on the many dimensions of the learning process.

Nevertheless, CORD's experience in working with thousands of teachers over a period of three decades is that the transition from conventional teaching to REACT-supportive teaching is possible, however gradual, and can be accomplished through a professional development *process* lasting several weeks.

REACT-supportive learning environments challenge students as well as teachers. Students are not asked to memorize lists of facts and dates or to manipulate data provided in textbooks. For any given problem, they are challenged to determine what data are required and to find and retrieve the data themselves. They learn to support and to depend on one another (as do adults in the workplace) and to challenge their own long-held assumptions. They are challenged to find connections that they had not formerly dreamed of and to take responsibility for their own learning.

While the challenges of REACT-supportive teaching and learning are great, so are the rewards. In his 2001 book titled *Contextual Teaching Works!*, Dale Parnell, one of the seminal figures in CTL, described several projects (U.S. Armed Services, Ford Foundation, CORD, U.S. Department of Education) that, in effect, tested the validity of REACT-style teaching. The results of all the studies cited agreed that contextual teaching has many benefits that conventional teaching lacks. Among those benefits are these:

1. Students try harder and are more interested in their studies.
2. Students behave better.
3. Absenteeism and tardiness are reduced.

4. Students enjoy contextual classes more than traditional classes.

5. In contextual classes, students are more willing to accept responsibility for their own learning.

6. Contextual teaching enables "slow learners" to achieve higher results than would otherwise be possible.

These findings are consistent with CORD's years of experience in working with teachers. Time and again, teachers who have worked through CORD's professional development process and have applied the REACT methodology in the classroom *over time* have expressed enthusiastic approval of the methodology. We close with these words from one of our teacher colleagues:

> Using this method [the REACT methodology] has really stretched me. I admit that. In a way it is harder, and it takes more time and I have had to "unlearn" some of my old habits. But it's worth it! The students make greater progress over the course of the school year. They take their projects seriously because the projects are related to things they consider important. I enjoy seeing the students get really excited, and my discipline problems have all but disappeared. My students continue to do well on their standardized tests. In fact, I think that as a result of contextual teaching, some of the slower learners actually do better than they would have if I had stuck to the old "sage on the stage" methods." (Catherine Williams, Midway High School, Hewitt, Texas)

The following are the other titles in the ABC series, along with a short description of how each addresses major life issues:

Air and Other Gases (ISBN 1-57837-087-6)   Students investigate gas exchange in plants and animals, the gas laws and their role in commercial applications of gases, and the roles of specific gases in problems such as smog and environmental degradation.

Animal Life Processes (ISBN 1-57837-073-6)   Students learn about homeostasis and how organisms maintain energy levels through digestion, absorption, and metabolism of nutrients; how fluid levels are maintained; how organisms produce and process waste; and how they regulate temperature.

Community of Life (ISBN 1-57837-079-5)   Students learn how different species of organisms depend on one another and how they adapt to climate and geography. The unit also shows how ecosystems change as a result of natural disasters such as fire and flood and human disturbances such as oil spills.

Continuity of Life (ISBN 1-55502-839-X)   Students learn how a knowledge of cell processes, genetics, and reproduction provides many practical benefits—in industry (greater efficiency in manufacturing and new products), in health care (new drugs and vaccines), and in society in general (reductions in crime and hunger).

Disease and Wellness (ISBN 1-57837-072-8)   Students explore the characteristics of diseases and immune system response as well as the precautionary steps that contribute to long-term wellness. The unit provides information and activities pertaining to the cultivation of behaviors, attitudes, and habits that are conducive to a high quality of life.

Microorganisms (ISBN 1-57837-081-7)   Students learn about both the positive and negative roles of microorganisms in our world. By acquainting students with the wide range of processes that involve microorganisms, the unit opens the door to many potential careers, including work in medical and research laboratories, food companies, public health facilities, wastewater treatment plants, and chemical plants, among others.

Natural Resources (ISBN 1-57837-085-x)   Students learn about how natural resources affect one another and human beings. Specifically, students explore fossil fuels, air, water, soil, and plants and animals and issues such as availability, depletion, and degradation. The unit also introduces students to occupations in which people take an active part in protecting and using natural resources.

Nutrition (ISBN 1-57837-077-9)   Students learn about diet, nutrition, and digestion in both people and livestock. The unit provides an introduction to the many occupations related to nutrition, from wheat farming to nutrition consulting and others of the many people who prepare, process, and safeguard food.

Plant Growth and Reproduction (ISBN 1-57837-079-5)   Students learn about what plants require to grow and about the many ways plants are used for food and fiber. The unit acquaints students with the pressures and difficulties faced by farmers and growers. It also provides career-related information about the many people who make their living processing plants.

Synthetic Materials (ISBN 1-57837-083-3)   Students learn about the chemical and mechanical properties of polymers, advanced ceramics, new metal alloys, and various composites of those materials for the purpose of understanding their uses in a wide range of products.

Waste and Waste Management (ISBN 1-57837-089-2) Students learn what "away" means when waste materials are "thrown away." The unit provides a broad range of opportunities for students to learn the reality and importance of garbage.

For more information on the ABC series, contact CORD Communications at 800-231-3015 or visit http://www.cordcommunications.com/.

REFERENCES

Bransford, J. D., A. L. Brown, and R. R. Cocking, eds. (1999) *How People Learn: Brain, Mind, Experience, and School.* Washington, DC: National Academy Press.

Crawford, M. (2001) *Teaching Contextually: Research, Rationale, and Techniques for Improving Student Motivation and Achievement in Mathematics and Science.* Waco, TX: CORD.

Harwell, S. (1999) *Why Do I Have to Learn This? Workbook.* Waco, TX: CCI Publishing, Inc.

Harwell, S. and W. E. Blank (2001) *Promising Practices for Contextual Learning.* Waco, TX: CORD.

Jensen, E. (1995) *Brain-Based Learning.* San Diego, CA: Turning Point.

National Commission on Mathematics and Science Teaching for the 21st Century (2000) *Before It's Too Late: A Report to the Nation from The National Commission on Mathematics and Science Teaching for the 21st Century.* Jessup, MD: Education Publications Center.

Parnell, D. (2001) *Contextual Teaching Works! Increasing Student Achievement.* Waco, TX: CORD.

Pintrich, P. R. and D. H. Schunk (1996) *Motivation in Education: Theory, Research, and Application.* Upper Saddle River, NJ: Prentice Hall.

CHAPTER 4

The Science of Terrorism: An Interdisciplinary Course for Nonscience Majors

LAURA POST EISEN

George Washington University, Washington, DC

In recent years, there have been a number of reports about the dismal state of scientific literacy in America (Goodstein 1992; Hazen and Trefil 1992). Although there are many reasons for wanting citizens to be scientifically literate (Goodstein 1992; Hazen and Trefil 1992; Angier 2008), one of the most compelling is that an increasing number of important public issues, such as stem cell research, genetic engineering, alternative energies, global warming, and teaching evolution, are deeply rooted in science. As a result, many of the choices that we all have to make as voters, consumers, and even parents necessitate some understanding of science. Furthermore, a scientifically literate public will be better able to resist the lure of antiscience, pseudoscience, and other forms of quackery. In their report *Science for All Americans*, the American Association for the Advancement of Science even makes the claim that "without a scientifically literate population, the outlook for a better world is not promising" (Rutherford and Ahlgren 1990).

Colleges and universities can help address America's scientific illiteracy by providing socially relevant courses for all nonscience majors (Hobson 2000). These courses provide an introduction to the nature of scientific inquiry that will benefit all students. However, some of our graduates will end up working at jobs that require them to have an even broader grasp of scientific issues. As the events of September 11, 2001 made clear, anyone who will be making or implementing policy related to national or international security will need to understand the nature of terrorist threats. Furthermore, those individuals who will be responsible for communicating about these same issues must be able to do so clearly and accurately. Thus, it is especially important for students who plan to work in the areas of international affairs, domestic politics, and journalism to understand the types of threats that we face, how we can best prepare for them, and what to do in the event of another terrorist attack. The Science of Terrorism course was designed to give these students the science background that they need in order to understand current terrorist threats, along with the conceptual framework to deal with future dangers. At the same time, the students develop skills that will help them to deal with these issues.

Nonscience majors are generally more receptive to studying science if they believe that the content is relevant to their own lives. Furthermore, students learn better when they are able to construct new knowledge from their own experiences (Herron 1996). Thus, the strategy of this course is twofold: to engage students with case studies based on terrorist incidents, so that they have a context for learning the science, and to provide a structure that encourages them to be active learners in the process. The cases are chosen in order to focus on specific ideas and chemical concepts, and are introduced with articles from newspapers or popular magazines. Once they are "hooked" by the story, the students, working in small groups, and with guidance from the instructor, make a list of the questions that the article raises, and the concepts and information that they need to know in order to understand the case. Each question is assigned to a group of students, who research the answers and report their findings to the rest of the class. In the discussions that follow, the instructor guides the students to discover the relevant concepts. Thus, the process follows the first steps of the learning cycle: students begin by exploring and engaging with the topic and go on to develop a concept. To complete the cycle, they later demonstrate their understanding by applying what they have learned to a new situation. As they go through this process, students develop teamwork and communication skills, along with a better understanding of the nature of scientific inquiry. Moreover, in contrast to many of their previous experiences with science classes, they discover that science is not only about "right" and "wrong" answers and that solving science problems may require reaching outside of strict disciplinary boundaries to use information from multiple fields. Finally, the students integrate all of these factors together by writing one or more policy papers in which they have to select one side of an issue and support their choice with evidence based on good science, as well as relevant information from other disciplines.

OVERVIEW OF THE COURSE

On the first day of the course, students are given a list of topics and are asked to identify which they would like to study during the semester. Not surprisingly, almost all choose poisons and explosions, while few, if any, select atoms or molecules. As chemists, we know that they cannot really understand poisons unless they know something about atoms and molecules, but we can take advantage of their interest in poisons to help them learn some broader chemical concepts. Thus, the course begins with five "elements of terror." As the students attempt to understand why each element is a terrorist threat, they learn about atomic structure and periodicity (from thallium), the atomic nucleus and radioactivity (polonium and cesium), chemical bonds and Lewis structures (chlorine), and chemical reactions (chlorine and phosphorus). Similarly, since explosions always fascinate students, chemical reactions, thermochemistry, and the first law of thermodynamics are introduced from the perspective of conventional explosives (ammonium nitrate/fuel oil mixtures and trinitrotoluene [TNT]). Although most explosions are strongly exothermic, a few (including triacetone triperoxide [TATP]) are actually entropy driven, which provides a connection to the second law of thermodynamics as well. Ultimately, however, it is the fear of nuclear, not conventional, explosions that concerns security experts, which leads to a discussion of nuclear fission, uranium enrichment, and mass–energy equivalence. Finally, bioterrorism is the motivation for exploring important concepts in biology and biochemistry. The structure and function of macromolecules is first introduced with an article about ricin poisoning, which leads to discussions about proteins and protein synthesis. However, ricin is not infectious: an attack with a microorganism such as smallpox or anthrax would present a far more serious danger. Viruses, including smallpox, are relatively simple, but in order to understand how they work, and the mechanism of action of most antiviral drugs, it is necessary to know about DNA structure and replication. In contrast, bacteria, such as anthrax, and the antibiotics that we rely on to treat bacterial infections, are much more complex, which provides opportunities to study other biochemical pathways as well. In both cases, the development of drug-resistant pathogens is a major medical problem that illustrates convincingly the process of evolution by natural selection. Of course, in the event of an actual bioterrorism incident, rapid identification of the infectious agent would be critical. Many methods are used for this purpose, giving students an opportunity to explore some of the techniques used in modern molecular biology. Unfortunately, there is nothing to stop terrorists from using these same methods, and the course concludes with the still hypothetical possibility of a terrorist attack using a genetically engineered microbe.

Part I: Elements of Terror

Thallium: An Introduction to Atomic Structure and Periodicity   The Science of Terrorism course takes an unusual approach to teaching about atoms, molecules, and reactions. Instead of the broad overview that is commonly used in introductory chemistry classes, the course begins with a close look at the chemistry of five interesting, diverse, and dangerous elements. The five elements are chosen in order to focus on different aspects of chemistry, and each is introduced with a news article that describes its use as an agent of terror. Thus, the first reading assignment is a newspaper article about the death of former Russian KGB agent Alexander Litvinenko. The article suggests that Mr. Litvinenko was the victim of thallium poisoning (Cowell 2006a). Of course, the article leaves many questions unanswered. What is thallium and where is it found? Why is it poisonous? How is thallium poisoning diagnosed, and can it be treated? What are the "advantages" and "disadvantages" of using thallium as opposed to other poisons? Each question that is identified can be assigned to a small group of students who will research the answer to present to the rest of the class. Since most students know little, if anything, about thallium beforehand, they immediately realize that this course will not just be a repeat of high school chemistry, and, after reading the article, most of them are interested in finding out more about this relatively unknown element.

Thallium has a long history as a real-life murder weapon (Emsley 2005) and has been implicated in a number of fictional crimes as well.[1]

---

[1] http://en.wikipedia.org/wiki/Thallium.

As a terrorist weapon, thallium sulfate was the "poison of choice" of Saddam Hussein's secret police in Iraq.[2] As we explore the chemistry of thallium, students who begin with limited knowledge of chemistry will learn the basics of atomic structure and periodicity, while those with strong chemistry backgrounds will be challenged to apply what they already know in order to understand an element that they undoubtedly did not study in high school. Since they are reading about poison, spies, and terrorists, both groups of students are likely to be more engaged than if they were studying atoms and elements in the abstract, while the nature of the problem makes it obvious that learning chemistry is relevant.

Thallium is named for the emerald green color that is seen when thallium salts are put in a flame. (Although students would undoubtedly love to see a live demonstration of the thallium flame test, it is unlikely that thallium salts will be available for this purpose. However, flame tests for other elements can be demonstrated instead, and the thallium emission spectrum can then be viewed online.[3]) When students investigate the methods that are used to diagnose thallium poisoning, they discover that the green emission line is related to the modern method of detecting thallium by atomic absorption spectroscopy. If time permits, they can explore for themselves how spectral lines can be used as an identification tool by observing a variety of atomic emission spectra with simple spectroscopes or diffraction gratings.

Of course, what the students are really interested in is why thallium is poisonous. Surprisingly, thallium is toxic because it mimics potassium in the body. But why would thallium behave like potassium? As we study the periodic table and chemical periodicity, there is no immediate reason to suspect that these two elements would have similar properties. A close look at the electron shell arrangement of thallium and potassium, however, reveals that both form +1 ions. Since Tl^+ ions also happen to be similar in size to K^+ ions, they are able to replace potassium ions in cellular processes. (Thallium poisoning is treated with a compound called Prussian blue, which binds to +1 ions and thus facilitates their removal from the body.) It is clear then that we cannot understand the toxicity of thallium without studying its atomic structure and electron distribution. But chemistry is only part of the story. The effects of thallium poisoning only make sense if the

---

[2]http://www.timesonline.co.uk/tol/news/world/middle_east/article642566.ece.
[3]http://webmineral.com/help/FlameTest.shtml.

physiological role of potassium in the human body is understood, which makes the interdisciplinary nature of the problem apparent to the students.

Finally, in order to expand this unit beyond thallium and to give students a chance to apply what they have learned about atomic structure, the class is divided into groups, and each group is assigned a different poisonous element. The students then explain the chemistry of "their" element to the rest of the class. (Since the class as a whole will be studying a few obvious candidates, including chlorine and phosphorous, later on, these elements would not be assigned at this point. There are, however, a number of interesting elements that can be explored, including arsenic, lead, and mercury [Emsley 2005].)

Polonium and Cesium: Exploring Nuclear Chemistry and Radioactivity   As it turns out, the early reports claiming that Mr. Litvinenko was poisoned with thallium were incorrect. In fact, a very different element was responsible. A second news article reveals that Mr. Litvinenko was actually the victim of radiation poisoning resulting from ingestion of polonium-210 (Cowell 2006b; McNeil 2006). This new information shifts our exploration of atomic structure to the nucleus. Once again, the students are asked to create a list of questions that serve as the framework for developing concepts, in this case related to nuclear chemistry. What is polonium-210 and why is it so deadly? Where is polonium normally found and how would a terrorist get access to it? What is radioactive decay and what happens to atoms when they decay? Is there a risk to other people who came in contact with Mr. Litvinenko after he ingested the poison? In answering these questions, the students learn about radioactivity, half-life, and balancing simple nuclear equations from the perspective of the isotope responsible for Mr. Litvinenko's death.

Once the basics of nuclear chemistry have been covered, we can move on to explore a major threat in the area of nuclear terrorism: the radiological dispersion device, or dirty bomb. Although any radioisotope could theoretically be incorporated into a dirty bomb, the U.S. government has identified nine isotopes that are most likely to be used in such a device (Table 4.1).[4]

---

[4] http://www.remm.nlm.gov/rdd.htm#isotopes.

TABLE 4.1   U.S. Government's List of Dirt Bomb-Making Radioisotopes

Isotope	Important Emission[a5]	Radiological Half-Life[6]	Biological Half-Life[7,8]	Chemically Mimics
Am-241	Alpha	430 years	50 years	—
Cf-252	Alpha	2.6 years	50 years	—
Cs-137	Beta, gamma	30 years	70 days	Potassium
Co-60	Beta, gamma	5.3 years	10 days	—
Ir-192	Beta, gamma	74 days	200 days	—
Pu-238	Alpha	88 years	50 years	—
Po-210	Alpha	140 days	60 days	—
Ra-226	Alpha, gamma	1600 years	44 years	Calcium
Sr-90	Beta	29 years	30–50 years	Calcium

[a]Includes emission by important decay products as well as by the isotope itself.

Since experts believe that cesium-137 is a prime candidate because of its wide availability,[9] we examine a hypothetical scenario in which a dirty bomb containing cesium-137 is detonated in Washington, DC (Anderson 2002). Any dirty bomb would certainly cause panic and disruption, but how would a bomb laced with cesium-137 differ from a bomb using a different isotope? One major factor is the type of radiation emitted. Thus, it is essential to understand how alpha, beta, and gamma radiation affect the human body. For example, cesium-137 emits both beta and gamma radiation, so that both internal and external exposure can be dangerous. It is also important to distinguish the biological half-life of an isotope (i.e., how long the isotope remains in the body) from its radiological half-life. A dirty bomb made with cesium-137 would require a major cleanup, since its radiological half-life is about 20 years. However, it is excreted relatively rapidly from the body, so that the health effects are limited to a much shorter time period. Finally, we examine the specific risks that occur when an isotope is chemically similar to one of the elements found naturally in the body. Cesium is an alkali metal and, as a result, behaves like potassium. Thus, exposure to cesium-137 presents the added danger associated with its tendency to accumulate where potassium is normally found. Of course,

[5]http://www.evs.anl.gov/pub/doc/ANL_ContaminantFactSheets_All_070418.pdf.
[6]Ibid.
[7]Ibid. (Am-241, Cf-252, Ir-192,Pu-238).
[8]http://hyperphysics.phy-astr.gsu.edu/Hbase/Nuclear/biohalf.html (Cs-137, Co-60, Po-210, Ra-226, Sr-90).
[9]http://www.wired.com/politics/law/news/2002/06/53110.

thallium also mimics potassium, and it is therefore not surprising that the same antidote (Prussian blue) is used to treat poisoning with radioactive cesium.[10] Similarly, strontium and radium isotopes are especially dangerous because they can replace calcium in bones and teeth. Once again, the students discover that they need to know about atomic structure and periodicity, along with some biology, in order to understand the health effects of an agent of terror. This unit concludes as students work in groups to research one of the other potential dirty bomb isotopes and then present their findings to the rest of the class, demonstrating again that they can apply what they have learned to a different example.

Chlorine: States of Matter and Chemical Reactions   Thallium and polonium are not the only chemical elements that have been used as terror agents. There are reports of sulfur being used by the Spartans to incapacitate Athenians during the Peloponnesian Wars in the fifth century BC, while in the fifteenth century, Leonardo DaVinci described how to use a powder containing arsenic to asphyxiate an enemy.[11] Our next case study focuses on a completely different element and a more recent episode in history. Chlorine was the first of the infamous poison gases used by the Germans during World War I (Irwin 1915). This element of terror has little in common with thallium and polonium, and different chemical concepts are required in order to understand its toxicity. In the first place, chlorine occurs naturally as a diatomic molecule, which provides a context for learning about chemical bonding and Lewis structures. Furthermore, unlike thallium and polonium, which are solids under standard conditions, chlorine is a gas, and when it comes to poisons, the difference between a solid and a gas is especially important. Although there are a number of factors that determine how toxic a substance is, the route of exposure is critical. Inhaled agents have the potential to affect many more people in the first place and generally act more rapidly than ingested substances once they enter the body.[12] Moreover, since chlorine gas is heavier than air, it does not disperse rapidly and therefore has more time to cause symptoms. Thus, in order to understand the toxicity of chlorine, we must also learn about states of matter and density. Finally, unlike thallium or polonium, chlorine is poisonous because of its chemical reactivity—

[10] http://www.fda.gov/cder/drug/infopage/prussian_blue/Q&A.htm.
[11] http://www.armageddononline.org/chemical_warfare.php.
[12] http://www.eoearth.org/article/Toxicity#Factors_Influencing_Toxicity.

specifically, it reacts with water to form hypochlorous and hydrochloric acids.[13] As a result, we need to know about chemical reactions, balancing equations, and acid–base chemistry in order to understand the nature of this chemical threat. Once again, the students can explore other poison gases, such as phosgene and mustard gas, as a way to apply what they have learned from studying chlorine.

Phosphorus: Investigating Chemical Reaction Energy We have now seen a number of different ways in which chemical elements can be toxic: Thallium replaces potassium in cellular processes; polonium-210 is radioactive and damages cellular components by emitting alpha particles; and chlorine reacts with water to form corrosive acids. Our final element of terror acts in yet another way. White phosphorus has a number of military uses—it is used to produce smoke screens and as an emergency light source. However, it also is a weapon that inflicts painful burns on unsuspecting victims (Emsley 2000; Frenkel and Evans 2009). Once again, the damage caused by phosphorus is related to its chemical reactivity. However, in this case, it is not the *product* of the reaction that matters. Instead, it is the *energy* change that occurs when the element reacts with oxygen. The oxidation of phosphorus ($P_4 + 5 O_2 \rightarrow P_4O_{10}$), like many other chemical reactions, releases energy—that is, it is exothermic. But where does the energy come from? Chemical reactions involve breaking existing bonds and forming new ones. Breaking bonds is always endothermic—that is, energy is always required to break a bond. On the other hand, forming new bonds is always exothermic. Thus, a reaction will be exothermic overall if more energy is released when the new bonds are formed than was required to break the existing bonds. Using Lewis structures, along with a balanced equation to represent the reaction, the number and types of bonds that must be broken and formed can be determined. The net energy change for the reaction can then be estimated using average bond enthalpies. (Although enthalpies of formation give the net reaction energies more accurately, using average bond enthalpies is conceptually much simpler, and makes it clear that it is the difference between the energy needed to break bonds and the energy released when new bonds are formed that determines whether a reaction is endothermic or exothermic overall.) Note that we are now considering the reaction of compounds, not elements, which provides an opportunity to explore chemical bonding and to use Lewis structures again.

---

[13] http://emedicine.medscape.com/article/820779-overview.

And, once again, we must be able to write and balance the equations representing chemical change.

Part II: Explosions and Energy

Conventional Explosions Most people do not know that the oxidation of phosphorous is what produces the light and heat when a match is lit. However, they are likely to be familiar with other combustion reactions that we all rely on every day. As most school children know, three things are required in order for combustion to proceed: a fuel, an oxidizer, and an ignition source (the familiar fire triangle). In the internal combustion engine, gasoline (the fuel) is mixed with air (the oxidizer) and is ignited with a spark. The resulting reaction releases energy, which causes the gaseous products to expand. The work done by the expanding gases is then coupled to the automobile drive train to move the car. Of course, the combustion of gasoline in our automobile engines is carefully controlled. Under different conditions, the gaseous products of a combustion reaction can expand violently, and an explosion results. Many explosions are just combustion reactions that take place very rapidly, so that the gaseous products expand out of control. We begin the unit on explosions with an article about the bombing of the Oklahoma City Federal Building (Johnston 1995). In this case, the explosive was a mixture of ammonium nitrate and fuel oil, known as ANFO. The rate of a combustion reaction depends on several factors, including the proximity of the fuel molecules to the oxidizer. Ordinarily, when a fuel burns in air, the reaction rate is limited by the availability of oxygen, since only 20% of the air molecules are reactive. Moreover, only the surface molecules are in contact with the oxidizer, so that just a small fraction of the fuel burns at a given time. One way to speed up the rate of a combustion reaction is to increase the surface area of the fuel. As a result, substances not ordinarily thought of as combustible can burn explosively; that is, flour in a grain elevator or finely powdered metals. Alternatively, the oxygen can be brought closer to the fuel. When ammonium nitrate is heated, pure O_2 is released: $2\ NH_4NO_3 \rightarrow 2\ N_2 + 4\ H_2O + O_2$. Thus, when ANFO is ignited, it is not just the surface molecules that are in contact with the oxidizer, since oxygen is now being produced within the explosive mixture. The heat that is released during the reaction further accelerates the decomposition, producing even more oxygen, and thus speeding up the reaction even more. Furthermore, the products of the decomposition of AN are gases, so that there is a rapid increase in volume. The rapid rate of the combustion reaction, combined with the production of large volumes of gas, is the recipe for an explosion.

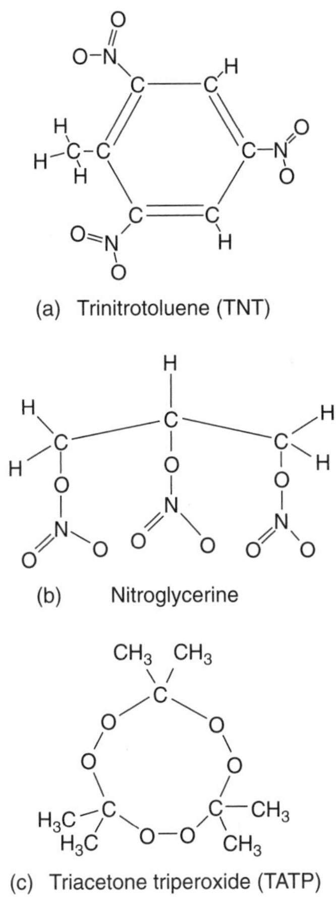

(a) Trinitrotoluene (TNT)

(b)   Nitroglycerine

(c) Triacetone triperoxide (TATP)

FIGURE 4.1   (a) Trinitrotoluene (TNT), (b) nitroglycerine, and (c) triacetone triperoxide (TATP).

Even though mixing an oxidizer and a fuel as in ANFO can obviously make an effective explosive, the two substances must be combined in just the right proportions for optimal results. Alternatively, instead of using separate molecules as the fuel and oxidizer, it is possible to combine these two functions in one compound. TNT is the standard by which explosives are usually measured. As can be seen in Fig. 4.1a, TNT combines a hydrocarbon component (the fuel) and nitrate groups (the oxidizers) in a single molecule. Thus, the combustible component is never far away from the oxygen that is produced when the nitrate groups are heated. Although TNT is a powerful explosive, it actually does not contain an optimal ratio of fuel to oxidizer. (Since there are seven carbon atoms per TNT molecule, it would require 14 oxygen atoms just to fully oxidize the carbons to CO_2.) Other molecular explosives, including nitroglycerine (Fig. 4.1b), have a "better" oxidizer-to-fuel ratio.

Entropy Also Matters Most explosions are driven by the favorable energy change that accompanies the reaction, but this is not always the case. In 2001, Richard Reid (the so-called shoe bomber) attempted to blow up an airplane using TATP (Fig. 4.1c) (Shenon and Bellock 2001). (It is this incident that led to restrictions on carrying liquids and gels onto airplanes.) TATP (also known as Mother of Satan) is a very unstable molecule that can be made by mixing hydrogen peroxide (not the drugstore variety, however), acetone, and an acid.[14] Note that TATP contains no nitrogen and is therefore not detected by screening devices designed to sense the more common nitrogen-containing explosive compounds. Surprisingly, it turns out that the explosion of TATP is not very favorable energetically. When TATP detonates, a relatively small amount of the material produces large volumes of gas in a fraction of a second;[15] the resulting entropy burst is what drives the reaction (Dubnikova et al. 2005). Thus, we have a context for learning about entropy and free energy. (It is also worth pointing out some more familiar examples of energetically unfavorable processes that nevertheless take place spontaneously. For example, melting of ice in a warm room and dissolving salts in an ice pack both represent spontaneous endothermic processes. Students are usually intrigued by the fact that most commercial ice packs contain the same AN that they studied as an explosive.)

Nuclear Explosions Although conventional explosives have become the weapons of choice of terrorist groups, a joint report issued in 2008 by Harvard's Kennedy School of Government and the Nuclear Threat Initiative reminds us that there is a real danger that terrorists could get and use a nuclear weapon.[16] In order to understand what this would mean, we return to the atomic nucleus. A nuclear fission reaction releases far more energy than any ordinary chemical process. The Oklahoma City bomb was equivalent to the explosion of approximately 4000 lb of TNT.[17] In contrast, the atomic bomb dropped on

[14] http://www.globalsecurity.org/military/systems/munitions/tatp.htm.
[15] Ibid.
[16] http://www.nti.org/e_research/STB08_Executive_Summary.pdf.
[17] http://www.fema.gov/rebuild/mat/mat_fema277.shtm.

Hiroshima exploded with energy equivalent to about 20,000 tons of TNT.[18] But where does all of this energy come from? Unlike ordinary chemical reactions, nuclear fission does not involve breaking and forming chemical bonds. Instead, the energy comes from the loss of mass that accompanies the fission reaction. Most, if not all, of the students will be familiar with Einstein's famous equation, $E = mc^2$, but few are likely to understand what it means.[19] In 1939, Lise Meitner and her nephew Robert Frisch reported their discovery of nuclear fission.[20] They realized that the energy that accompanied the fission of uranium nuclei could be accounted for by using Einstein's equation.

Nuclear fission occurs when a heavy nucleus is so destabilized by the capture of a neutron that it splits into two smaller nuclei. Only a few isotopes, including uranium-235 and plutonium-239, are capable of undergoing fission. When the nucleus splits, additional neutrons are released as products. If these neutrons hit other fissionable nuclei, a chain reaction can occur. If, on the other hand, most of the neutrons strike nonfissionable objects, the chain reaction is interrupted. Although uranium-235 is fissionable, the more common uranium-238 isotope is not. Since naturally occurring uranium contains over 99% uranium-238, there is not enough uranium-235 present for even a low-level chain reaction to propagate. Thus, in order to have any kind of chain reaction, the uranium must be enriched in order to increase the ratio of uranium-235 to uranium-238. The degree of enrichment needed depends on how the energy produced from the fission is going to be used: for a nuclear power plant, it is sufficient to have about 3–5% uranium-235, whereas for a weapon, the percentage must be significantly higher.[21] Fortunately, separating uranium-235 from uranium-238 is not easy and requires highly specialized equipment. Since it is not possible to produce a nuclear weapon without enriched uranium, international watchdog groups are constantly looking for evidence that a terrorist group or a rogue nation is trying to assemble an enrichment facility. In fact, Iran's decision to restart its enrichment program provoked an international crisis in 2006 (Moore 2006).

[18] http://avalon.law.yale.edu/20th_century/mp11.asp.
[19] The NOVA video "Einstein's Big Idea" does an excellent job of explaining the history and meaning of the equation.
[20] http://www.atomicarchive.com/Docs/Begin/Nature_Meitner.shtml.
[21] http://www.nrc.gov/reading-rm/doc-collections/fact-sheets/enrichment.html.

Part III: Bioterrorism and Biochemistry

Ricin and Proteins   The threat of a nuclear attack is one reason that Congress established the bipartisan Commission on the Prevention of Weapons of Mass Destruction Proliferation and Terrorism in 2007. In December 2008, the Commission issued "World at Risk," a report that cautions that a weapon of mass destruction is likely to be used somewhere in the world before the end of 2013 unless action is taken immediately to forestall such an event.[22] There are two major categories of weapons about which they are most concerned. Obviously, nuclear weapons in the hands of terrorists could kill enormous numbers of people, and the Commission agrees that the risk of a nuclear terrorist attack is increasing. However, they predict that it is more likely that terrorists will obtain and use a biological weapon first.[23]

We begin our exploration of biological terrorism with an incident that sounds as if it came straight out of a James Bond movie. In 1978, Georgi Markov, a Bulgarian novelist and dissident, was walking across a bridge in London when he felt a pain in his thigh (Edwards 2008). He turned around and noticed a man picking up an umbrella. Markov continued on to work that day, but he became very ill and had to be hospitalized in the evening. Doctors were unable to determine the cause of his illness, but when he died 3 days later, a small pellet was discovered embedded in his leg. Apparently, the pellet was fired at Mr. Markov from the tip of the umbrella carried by the unidentified assassin. Although no poison was ever found, forensic tests concluded that Mr. Markov was killed with the poison ricin. More recently, there have been a number of episodes in which ricin has been linked to possible terrorist and criminal activity. In 2003, there were reports that ricin had been found in a London apartment used by alleged Al Qaeda operatives. This led to much speculation about a possible link between the group and Iraqi extremists, and the Bush administration used this supposed poison cell to support the U.S. invasion of Iraq (Pincus 2005). (In the end, it turned out that there was no ricin, or any other poison, in the apartment.) In 2004, three U.S. Senate office buildings were closed after ricin was found in letters addressed to Senate Majority Leader Bill Frist (Johnston and Hulse 2004). After the Senate incident, it was revealed that letters containing ricin had been mailed to the White House a few months earlier. Fortunately, no one was poisoned in either case, and it is believed that these letters were the work of a

---

[22]http://www.preventwmd.gov/report/.
[23]Ibid.

lone individual who was unhappy about a particular piece of legislation (Jacobs 2004). In 2008, a man was admitted to a hospital in Las Vegas complaining of difficulty breathing. When his hotel room was searched, vials of crude ricin were found (Freiss 2008). The man survived, but investigators have not revealed what, if anything, he planned to do with the ricin. The mere threat of ricin has also been used to scare people. In January 2009, 11 gay bars in Seattle received letters warning that customers would be poisoned with ricin (Perry 2009), although once again no one was hurt.

Ricin is a protein that is extracted from common castor beans. Although as little as 500 µg, which could fit on the head of a pin, is enough to kill a person, most experts feel that for technical reasons, it is not likely to be used as a weapon of mass destruction (Shea and Gottron 2004). However, since castor beans are relatively easy to find and grow, ricin is considered to be a serious small-scale terror threat, and for this reason, it is listed as a Category B biological agent (the second-highest priority) by the Centers of Disease Control.[24] Ricin acts by inhibiting protein synthesis, and in order to understand the toxicity of this compound, it is necessary to know something about proteins and how they are made in cells. This provides a context for learning about one aspect of the "central dogma" of molecular biology: DNA → RNA → protein. If desired, students can explore other "molecules of murder" (Emsley 2008) as a way to learn about additional cellular mechanisms.

Smallpox and Anthrax: DNA and Antimicrobial Drugs   Although Category "B" threats, including ricin, must be taken seriously, the major focus of public health preparedness has to be on Category "A" agents, which have the potential to cause casualties rivaled only by nuclear weapons. Richard Preston's real-life thriller *The Demon in the Freezer* (Preston 2002) examines the history of man's struggle against the smallpox virus, one of the Category A threats, and considers the possibility that the virus could be used by bioterrorists. Although a smallpox attack remains hypothetical, Preston also investigates the 2001 anthrax letters, which serve to remind us that bioterrorism is a very real threat. (Anthrax is also on the Category A list.) Since smallpox is a viral disease, while anthrax is transmitted by bacteria, the book is a good introduction to both types of infectious agents. There are many interesting aspects about viruses and bacteria that can be studied if time

[24] http://emergency.cdc.gov/publications/feb08phprep/appendix/appendix6.asp.

permits, but focusing on the mechanism of action of antiviral and antibacterial drugs is especially relevant for a chemistry class. Viruses replicate by taking over host cell functions, which makes it is difficult to find drugs that interfere with viral growth without affecting the host cell as well. As a result, there are only a small number of effective antiviral compounds, most of which rely on inhibiting viral DNA synthesis. In order to understand how these drugs work, it is necessary to know something about the structure of DNA, as well as how DNA is replicated and transcribed. Thus, we have a meaningful context to study another aspect of the central dogma. In contrast, since bacteria must be able to survive without a host cell, they utilize a number of biochemical pathways that are not found in human cells. Each of these pathways is a suitable target for antibacterial drugs, and as a result, the mechanism of action of antibiotics is quite varied. Unfortunately, it is becoming clear that both viruses and bacteria can eventually develop resistance to most, if not all, of these drugs, a very real and frightening example of evolution by natural selection in action (Antonovics et al. 2007).

The threat of a bioterrorist attack with smallpox is especially disturbing since the eradication of smallpox remains one of the great achievements in human history. Unfortunately, since routine vaccination against smallpox was discontinued in 1978, few people retain immunity today. Although there are only two official repositories of the smallpox virus today, it is still possible that terrorists will be able to obtain the virus. Thus, the government has had to stockpile supplies of the vaccine, and there is some debate about whether to resume routine smallpox vaccinations. Although the smallpox vaccine was discovered by accident, the story of how Louis Pasteur developed the first anthrax vaccine and his use of unvaccinated animals as controls remain as excellent lessons about the process of science (Trachtman 2002).

As we learned after the anthrax attacks in 2001, the ability to rapidly detect and to identify a bioterrorism agent is critical. A variety of methods are used for this purpose, including DNA fingerprinting, DNA sequencing, PCR, and ELISA immunoassays. As students read about how these methods are utilized in fighting terrorism, they learn how the methods work. It is even better if they can actually apply at least some of the methods in the laboratory. A number of suppliers provide kits designed for educational use,[25] and in most cases, it is possible to use these activities in a way that simulates a bioterrorism scenario.

Genetically Engineered Biological Weapons Unfortunately, terrorists can also use many of the methods that were originally developed

[25]See, for example, http://www.biorad.com/ and http://www.edvotek.com/.

for scientific research. Scientists now have the ability to synthesize viruses from scratch (Cello et al. 2002) and to manipulate genes to engineer new strains of infectious agents. "World at Risk" warns that as genetic engineering and other sophisticated technologies become more widespread, the likelihood that terrorists will use these methods to create a biological weapon increases.[26] Thus, the ultimate bioterrorist threat may be a new strain of bacteria or virus engineered so that it is very contagious, has a high mortality rate, and resists existing vaccines and drugs. Fortunately, we have not had to deal with such a catastrophe yet. However, Richard Preston gives us a preview of what such an event might be like. *The Cobra Event* (Preston 1997) describes a fictional terrorist who constructs a deadly weapon using genes from three real viruses. Preston himself, in an afterword to the novel, says that although the book is a work of fiction, "The non-fiction roots of the book run deep … the historical background is real, the government structures are real, and the science is real or based on what is possible." In fact, according to *The New York Times* (Miller and Broad 1998), when President Clinton read the book, he was so alarmed that he asked experts to assess whether the book described a realistic threat and went on to increase spending for bioterrorism preparedness as a result. Students can get hands-on experience with some of the methods of genetic engineering by using one of the kits that enable them to transfer a gene into bacteria, then to express the gene and isolate the corresponding protein.[27]

EPILOGUE

The afterword to *The Cobra Event* concludes with a quote from Thucydides: "Hope is an expensive commodity. It makes better sense to be prepared." But what is the best way to be prepared for a terrorist attack? Certainly, science education is a major factor. We need workers and citizens who have the knowledge to deal with known threats, and also the background to help them cope with as-yet-unknown possibilities. Although this course focuses on the science needed to understand specific types of chemical, nuclear, and biological terrorism, it also exposes students to some important general ideas and concepts in the natural sciences. In his book *Galileo's Finger: The Ten Great Ideas*

---

[26] Op. cit. http://www.preventwmd.gov/report/.
[27] For example, students can transfer the gene for green fluorescent protein into *Escherichia coli*, and then isolate the protein using chromatographic separation. See http://www.bio-rad.com/.

of Science, Peter Atkins identifies his candidates for the 10 biggest ideas of science (Atkins 2004). Students taking this course will be introduced to at least five of Atkins' choices: Matter is atomic; energy is conserved; change leads to disorder (entropy); inheritance is encoded in DNA; and evolution proceeds by natural selection. Thus, they will have a better understanding of how the natural world works, and will be better prepared to make informed and appropriate decisions when needed.

REFERENCES

Anderson, P. (2002) *Radiological Dispersal Devices: The "Dirty Bomb" Challenge.* Washington, DC: Center for Strategic and International Studies.

Angier, N. (2008) *The Canon.* New York: Houghton Mifflin.

Antonovics, J., J. L. Abbate, C. H. Baker, D. Daley, M. E. Hood, C. E. Jenkins, L. J. Johnson, J. J. Murray, V. Panjeti, V. H. W. Rudolf, D. Sloan, and J. Vondrasek (2007) Evolution by any other name: Antibiotic resistance and avoidance of the e-word. *PLoS Biology 5*(2), e30. doi:10.1371/journal.pbio.0050030

Atkins, P. (2004) *Galileo's Finger: The Ten Great Ideas of Science.* New York: Oxford University Press.

Cello, J., A. V. Paul, and E. Wimmer (2002) Chemical synthesis of poliovirus cDNA: Generation of infectious virus in the absence of natural template. *Science 297,* 1016–1018.

Cowell, A. (2006a) Intrigue swirls in ex-KGB man's illness. *New York Times,* November 20, p. A6.

Cowell, A. (2006b) London riddle: A Russian spy, a lethal dose. *New York Times,* November 25, p. A1.

Dubnikova, F., R. Kosloff, J. Almog, Y. Zeiri, R. Boese, H. Itzhaky, A. Alt, and E. Keinan (2005). Decomposition of triacetone triperoxide is an entropic explosion. *JACS 127,* 1146–1159.

Edwards, R. (2008) Poison-tip umbrella assassination of Georgi Markov reinvestigated. http://www.telegraph.co.uk/news/2158765/Poison-tip-umbrella-assassination-of-Georgi-Markov-reinvestigated.html (accessed November 28, 2009).

Emsley, J. (2000) *The Sordid Tale of Murder, Fire, and Phosphorus.* New York: Wiley.

Emsley, J. (2005) *The Elements of Murder.* New York: Oxford University Press.

Emsley, J. (2008) *Molecules of Murder: Criminal Molecules and Classic Cases.* Cambridge, U.K.: Royal Society of Chemistry.

Frenkel, S. and M. Evans (2009) Israel rains fire on Gaza with phosphorus shells. Times Online. http://www.timesonline.co.uk/tol/news/world/middle_east/article5447590.ece (accessed November 28, 2009).

Friess, S. (2008) Vials of ricin are found in Las Vegas hotel: Man is hospitalized. *New York Times*, March 1, p. A9.

Goodstein, D. (1992) The science literacy gap. *Journal of Science Education and Technology 1*, 149–155.

Hazen, R. and J. Trefil (1992) *Science Matters: Achieving Science Literacy*. New York: Anchor Books.

Herron, D. (1996) *The Chemistry Classroom*. Washington, DC: American Chemical Society, pp. 43, 56–57, 252–256.

Hobson, A. (2000) Teaching relevant science for science literacy. *Journal of College Science Teaching* Dec., 238–243.

Irwin, W. (1915) Germans use blinding gas to aid poison fumes. *New York Tribune*, April 27, p. 1. http://www.lib.byu.edu/index.php/The_Use_of_Poison_Gas (accessed November 28, 2009).

Jacobs, A. (2004) Truckers look in their ranks for "Fallen Angel" writer. *New York Times*, February 5, p. A28.

Johnston, D. (1995) Terror in Oklahoma City. *New York Times*, April 20, p. A1.

Johnston, D. and C. Hulse (2004) Concern over ricin discovery disrupts senate activity. *New York Times*, February 4, p. A1.

McNeil, D. (2006) A rare material and a surprising weapon. *New York Times*, November 25, p. A9.

Miller, J. and W. Broad (1998) Exercise finds U.S. unable to handle germ war threat. *New York Times*, April 26, p. A1.

Moore, M. (2006) Iran restarts uranium program. *Washington Post*, February 14, p. A1.

Perry, N. (2009) 11 gay bars get letters threatening ricin attacks. *Seattle Times*, January 9, http://seattletimes.newsource.com/html/localnews/2008597989_ricinthreat07m.html

Pincus, W. (2005) London ricin finding called a false positive. *Washington Post*, April 14, p. A22.

Preston, R. (1997) *The Cobra Event*. New York: Ballantine.

Preston, R. (2002) *The Demon in the Freezer*. New York: Random House.

Rutherford, F. J. and A. Ahlgren (1990) *Science for All Americans*. New York: Oxford University Press.

Shea, D. and F. Gottron (2004) Ricin: Technical background and potential role in terrorism. Congressional Research Service. http://www.fas.org/irp/crs/RS21383.pdf (accessed November 28, 2009).

Shenon, P. and P. Bellock (2001) FBI tests find explosive in shoes of jet passenger. *New York Times*, December 24, p. A1.

Trachtman, P. (2002) Hero for our time. *Smithsonian Magazine*, January, pp. 34–41.

CHAPTER 5

Chemistry for the Twenty-First Century: Bringing the "Real World" into the Lab

GAUTAM BHATTACHARYYA

Clemson University, Clemson, SC

INTRODUCTION

Concrete examples that help connect students' lived experiences to abstract scientific concepts are a popular means by which educators attempt to engage students. It is hoped that these "real-world" applications will help and motivate students to learn scientific concepts and even to consider becoming practicing scientists. Interestingly, these "real-world" examples are often used as entertaining asides rather than as the focus of the concept-building process.

As a lecturer in the Department of Chemistry at the University of Oregon, I had the privilege to participate in the ground-breaking program in green organic chemistry developed by Profs. K. Doxsee and J. Hutchison (2004). While teaching these courses, I quickly discovered the breadth of "real-world" topics I was able to discuss as an integral part of the course, not just as interesting asides. As illustrated in Fig. 5.1, teaching organic chemistry from a "green perspective" allowed me to explicitly connect the course material with a wide variety of other disciplines such as industrial chemistry, molecular biology, and toxicology. This was one of the first experiences I had in which I could

Making Chemistry Relevant: Strategies for Including All Students in A Learner-Sensitive Classroom Environment, Edited by Sharmistha Basu-Dutt
Copyright © 2010 John Wiley & Sons, Inc.

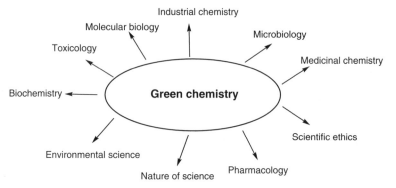

FIGURE 5.1   A sampling of disciplines that can be presented in a green chemistry discussion.

explicitly teach multiple sciences together, providing a truly interdisciplinary experience.

CONTEXT

Ever since the public became aware of the high-profile incidences of environmental pollution in the late 1960s through the 1970s, chemists, especially those in industrial settings, directed significant attention to developing more ecologically friendly processes. More recently, Paul Anastas and John Warner proposed a set of 12 principles that came to be known as green chemistry (1998). These principles can be summarized by the following philosophy:

- preventing the formation of waste,
- employing safer reagents and solvents,
- implementing selective and efficient transformations, and
- avoiding unnecessary transformations (Doxsee and Hutchison 2004, p. 51)

Doxsee and Hutchison (2004, p. 51) further conceptualize risk of adverse effects from chemicals as a function of intrinsic hazard and exposure, which they summarize in the following equation:

$$\text{Risk} = f(\text{exposure}, \text{hazard})$$

They argue that the traditional method of reducing risk is to minimize exposure, while the green philosophy is to minimize the intrinsic toxicity of a substance. The ideas behind greener chemical methods are not to perpetuate a fear of chemicals in the students' minds. Rather, they are to help students understand that, when used in an environmentally and ecologically conscientious manner, chemicals and chemistry produce many of the staples of human existence. Green chemistry has taken a center stage in the new millennium as the sustainability of the human race has become the most important scientific, economic, social, and political challenge in the world.

It is important to note that "green" is not a destination; it is a process. There can never be a perfectly green process, just one that is greener than the existing one. (The term "brown" is often used as an antonym of green when describing chemical processes.) The relativistic aspect of the concept of greener science is particularly useful to challenge college students, who tend to see things in more absolutist terms (Perry 1970), with ideas that have varying shades of gray. From the standpoint of teaching organic chemistry, these points are a perfect platform on which to discuss the research challenges at the frontier of the field. Thus, the key is that these principles are not limited to green chemistry but are the basis of good organic chemistry.

Although it is easy to offer students sobering cases of chemical mishaps, those instances tend to be meaningless, abstract facts in the minds of students. Conceptualizing the magnitude and nature of the problem is not a trivial matter and involves understanding why chemistry may be needed at all. From the viewpoint of a former synthetic organic chemist, I suggested to my students that chemistry can be used to prepare substances that are sparsely available from nature or not made by nature at all. These substances often exhibit important pharmacological, electronic, mechanical, or other properties that people find useful. One example of this idea that I offered was adipic acid, which the students synthesized during one of their experiments (see Experiment 5 in Doxsee and Hutchison 2004). Once the connection between adipic acid and nylon is made and the number of commonly used items made from nylon is discussed, it becomes easier to comprehend the need for a mass-scale synthesis. Since adipic acid is a minor oxidation product of fat, it would be impossible to meet the world's demands for nylon by extracting adipic acid from a natural source. Chemists, therefore, had to develop methods of synthesizing it so that a large percentage of the world's population could access products made from nylon. However, students have difficulty in fathoming the magnitude of waste that accompanies the global-scale production of chemicals such as adipic acid.

To help students put the magnitude of waste production in perspective, I offered a "back-of-the-envelope calculation" of a smaller-scale scenario based on a medicine that at least some students have used, Augmentin™. Augmentin is an antibiotic used to treat aggressive infections, particularly those that are penicillin resistant (Katzung 1995). A typical dose of the drug contains 875 mg of amoxicillin and 125 mg of potassium clavulanate, a potent beta-lactamase inhibitor. A 10-day course of two tablets per day is prescribed to adults with severe infections. Thus, one course of Augmentin requires 20 g of drug per adult. If everyone in a class of 250 students needed it that would add up to 5000 g, or 5 kg. For a University of 20,000 students, faculty, and staff this amounts to 400 kg of Augmentin for a single episode! How much would it take for a city or country?

For each gram of drug, if 1 g of waste is produced, then students can begin to imagine the magnitudes that need to be considered, assuming any one of us can actually fathom what a million, or billion, kilograms looks like. I finish by posing the question: Even if the waste were pure water, what effect would it have on the environment?

Although there were many ways of organizing the material, I chose a sequence that was influenced by my training as a synthetic organic chemist. Accordingly, the topics were arranged based on the components of chemical reactions:

$$\text{Starting material(s)} \xrightarrow[\substack{\text{Solvent(s)}\\\text{Energy}}]{\text{Reagent(s)}} \text{Product(s)} \xrightarrow{\text{Purification}} \text{Pure product(s)}$$

What this brief schematic of a chemical transformation helps convey is that to get the desired product(s), we must invest to buy or to use each one of the components. However, we also need to invest even more to dispose all the materials used other than the desired product(s). However, the cost of disposing the waste is not the biggest concern, since even with all the safeguards in waste disposal, there are still harmful effects that we cannot always control.

FEEDSTOCKS—THE STARTING MATERIALS OF CHEMISTRY

Where do reactants come from? For such a simple question, the answer is particularly fascinating. Typical training in synthetic organic chemistry, even at the doctoral level, does not address this issue explicitly. In fact, graduate students frequently invoke the phrase "Aldrich synthesis" without considering where Aldrich gets its materials. This term is the equivalent of believing that the source of physical constants is the appendix of a textbook (G. Bodner, pers. comm.).

The vast majority of the top 50 organic chemicals produced in the United States, according to data from *Chemical and Engineering News*

(ACS 2008) are derived from petrochemicals. Students are shocked to find, in that light, they wear petroleum, walk on petroleum, eat petroleum, and so forth. For example, adipic acid is primarily made by oxidizing cyclohexene, which is, in turn, made from benzene, a petrochemical isolated from the distillate of cracking of crude oil (Ophardt 2003). In recent years, this fact has gained special significance since the price of oil has played a significant role in the global economical downturn. Although the focus of the media discussion tends to be on the use of oil as a fuel, the chemical industry is equally, if not more, affected by the value of oil as the source of feedstocks. This fact helps explain the deep geopolitical influence that the price of crude oil wields; every facet of our material existence is derived from it. It also reshapes the whole debate about future petrochemical resources for the students.

Doxsee and Hutchison (2004) suggest several greener alternatives to petrochemicals. Two of them, microorganisms and biomass feedstocks, are particularly appropriate for organic chemistry courses, since most of the students' interests are geared toward the biomedical sciences. One example that has garnered much attention from the media over the past year is ethanol, which used to be produced from ethylene, a petrochemical (Tokay 2005). However, most of the world's ethanol production today is from the fermentation of starch using yeast. In Brazil, the source of starch is sugarcane molasses, but it is corn in the United States. The production of ethanol from corn has raised many ethical questions since the emergence of ethanol as a fuel increased the value of corn. Higher prices of corn made it less accessible to people in less-developed economies, which, in turn, is believed to have contributed to their ongoing problems with malnutrition.

The discussion of using microorganisms to produce feedstocks can include key techniques of molecular biology, since the advances in recombinant DNA technology has been the driver of this alternative. In addition to the better-known polymerase chain reaction (PCR) and site-directed mutagenesis, Cre/*lox* recombination helped revolutionize this industry (Kuhn and Torres 2002). In this technique, the enzyme Cre (cyclization recombination) attaches strands of DNA in bacteriophage *P1* at specific sites called *loxP*. This method allows scientists to delete entire genes and to recombine phage DNA, opening an entire world of organisms with mutations and gene knockouts. The Cre/*lox* story is a beautiful example of how a development in one field can impact several others.

The second greener alternative that can resonate with the organic chemistry audience is biomass feedstocks, chemicals that are directly extracted from renewable sources (Tokay 2005). Biomass feedstocks

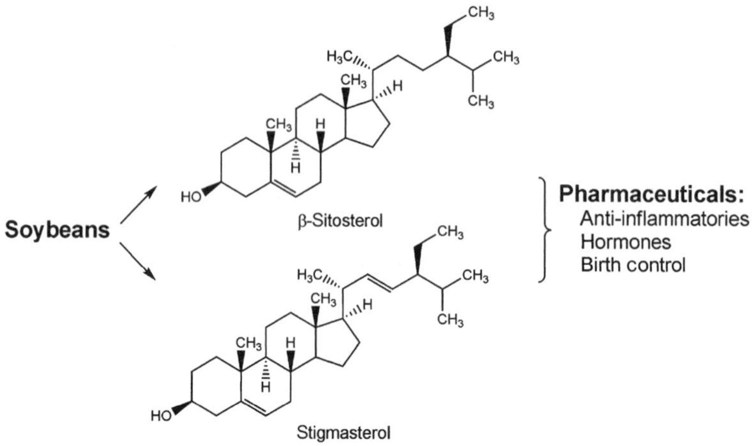

FIGURE 5.2 Pharmaceutical precursors isolated from soybeans.

have diverse industrial applications from essential oils used in food or health and beauty items to pharmaceuticals. One of the common uses of biomass feedstocks in the pharmaceutical industry is in the semisynthesis of complex compounds containing several stereocenters. In this process, an advanced intermediate, often containing all of the necessary stereocenters, is isolated from a renewable source and is converted into the desired drug. The classic examples are the conversion of stimasterol and β-sitosterol—both of which can be isolated from soybeans—into the steroid hormones or corticosteroids (Fig. 5.2). As inflammatory responses are linked to more and more medical conditions, the use and importance of corticosteroids is ever increasing. Another example is the synthesis of paclitaxel (Taxol™) from 10-deacetylbaccatin III (10-DAB) (Fig. 5.3). Isolated from the bark of the Pacific yew tree, Taxol is such a minor component that it is estimated that 14 kg of bark is required to produce enough drug to treat one patient for 1 day (Wei 2007)! Furthermore, due to the complexity of the structure, the total synthesis was not a viable option either. Thus, Holton and coworkers devised a semisynthetic route that was commercialized by Bristol-Meyers Squibb. In this process, 10-DAB, which is abundantly available from the needles of the more common English yew tree, is converted to Taxol in about 80% overall yield (Holton et al. 1994).

The discussion of biomass feedstocks can also include an examination of ethical and moral dilemmas, since supplying adequate amounts of raw materials can deplete renewable resources like water for irrigation or nutrients in soil. Furthermore, ensuring high crop yields can

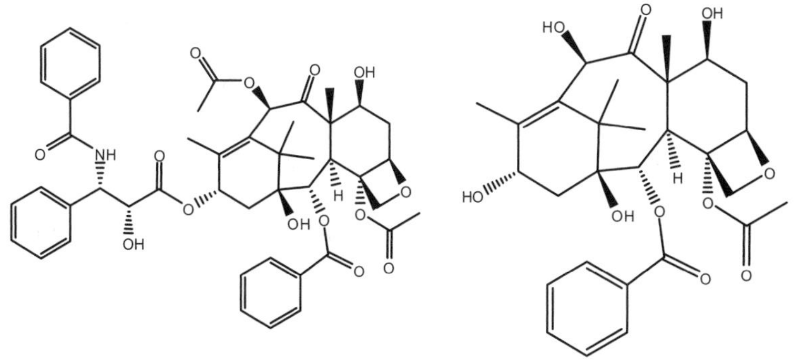

Paclitaxel (taxol) 10-Deacetylbaccatin III (10-DAB)

FIGURE 5.3 One of the most potent chemotherapeutic agents used in cancer, Taxol is semisynthesized from 10-DAB. 10-DAB is isolated from the needles of the English yew or is produced by a plant cell fermentation process.

require the use of fertilizers and pesticides that can be highly toxic. Again, "the lack of a clear-cut alternative, that is one without adverse side effects," can promote the students' ability to work with ambiguous situations.

REAGENTS

This topic can offer powerful ways of connecting the principles of chemical reactivity with those of greener chemistry. For example, a major goal of using greener processes is to minimize waste. In a chemical context, this means minimizing the formation of by-products, since they cause the greatest amount of waste formation. Not only must the unwanted substances must be disposed but solvents and/or energy must also be expended to separate the desired product from the undesired ones. These solvents constitute the largest fraction of the waste. Although, strictly speaking, catalysts are not reagents, their use and development are included in this discussion as alternatives to stoichiometric reagents.

Eliminating the formation of by-products involves developing reagents (catalysts) that selectively perform the desired transformation. A chemist's ultimate goal is to convert the starting material(s) into the desired product(s) using catalytic amounts of another substance. This goal is, perhaps, the single greatest challenge for organic chemists since it requires a fundamental understanding of how molecules react in solution. Practicing organic chemists rely on their understanding of the reaction's mechanism to make educated and rational choices when optimizing reaction conditions. Our research on how students solve mechanism tasks has shown that even first-semester graduate students treat mechanisms as blueprints for reactions that must be memorized (Bhattacharyya and Bodner 2005). They cannot see any use for understanding reaction mechanisms. Thinking about selective reagents, therefore, can provide a purpose for mechanisms beyond points on exams.

A second major issue that is related to the overall discussion of reagents is the practical considerations of performing chemical reactions. Past research has shown that even early-career organic chemistry graduate students do not take experimental plausibility into account when solving organic synthesis tasks (Bowen 1990; Bowen and Bodner 1991). For example, students will propose steps in which Grignard reagents are added to formaldehyde or to ethylene oxide. Although these reactions are technically plausible, practicing chemists rarely, if ever, will ever perform them due to the practical difficulties associated with using the reagents. Both ethylene oxide (b.pt.: 10.7 °C) and formaldehyde (b.pt.: −19.3 °C) are gases at room temperature. Conducting reactions with these reagents, therefore, requires a lot of energy to keep the reaction cooled and also requires a lot of manipulations to ensure purity of the gaseous reagents. Thus, the expenditure of energy to successfully carry out this reaction makes them highly inefficient. The applications of mechanistic organic chemistry and practical experimental considerations, therefore, can take "center stage" in discussions of greener alternatives to traditional organic reagents.

Biological Organisms

Biocatalysis typically involves a single biological molecule, an enzyme or RNA. Organisms, in contrast, refer to the use of the machinery of the entire organism. The production of ethanol discussed in the section on alternative feedstocks is an example of an organismal "reagent." The use of biological organisms is not limited to feedstock materials; rather, they are used for several important end-market products. For example, under a different set of fermentation conditions, the yeast that is used to generate ethanol, *Saccharomyces cerevisiae*, is used to make glycerol, which is used in liquid soaps, cosmetics, and in many other items (Chotani et al. 2000). As another example, Bristol-Meyers Squibb received the 2004 Greener Synthetic Pathways Award from the United States Environmental Protection Agency (EPA) for their development of a plant cell fermentation process for making 10-DAB.

Oxidizing/Reducing Agents

In traditional treatments of introductory organic chemistry, the paradigmatic oxidizing agents for alcohols and/or aldehydes are derived from Cr(VI). The first of these reagents combined Cr(VI) compounds with strong acids. The Jones reagent is one such example in which CrO_3 is mixed with concentrated sulfuric acid, producing chromic acid (Bowden et al. 1946). A major drawback of the Jones reagent is that primary alcohols are oxidized all the way to carboxylic acids; the aqueous environment makes it impossible to stop oxidation at the aldehyde stage. The harsh acidic environment constitutes a second disadvantage, since many functional groups and structural motifs are unable to survive the reaction conditions. Pyridinium dichromate (PDC, also known as Cornforth reagent) (Cornforth et al. 1962), Collins reagent (Collins et al. 1968), and pyridinium chlorochromate (PCC) (Corey and Suggs 1975) were created to, in part, circumvent these problems. Although these reagents were some of the most useful oxidizing agents available, they fell out of favor as the extraordinary toxicity of Cr(VI) became better known.

Over time, several alternatives to Cr(VI) oxidizing agents have been developed. However, each one of these presents major challenges to greener principles. For example, the activated-dimethylsulfoxide (DMSO) oxidations produce dimethylsulfide as a by-product, which is highly toxic. Additionally, the most widely used of reactions in this class, the Swern oxidation (Omura and Swern 1978), also produces carbon monoxide. As another example, Dess–Martin periodinane (Dess and Martin 1983) is made through an intermediate that can be highly explosive. Finally, one of the more benign oxidants, N-methylmorpholine-N-oxide (NMO), used in conjunction with tetra-n-propylammonium perrhuthenate (TPAP) as a catalyst (Griffith et al. 1987), produces N-methylmorpholine as the by-product, which results in the loss of a large amount of molecular weight that ends up in the waste stream.

In addition to these examples, nitric acid is one of the most commonly used oxidizing agents in industrial settings. It is used, for example, in the synthesis of adipic acid from cyclohexene (Doxsee and Hutchison 2004). Noyori and coworkers note that this process accounts for more than 10% of the NO_x gases produced by humans worldwide (Sato et al. 1998). The previously mentioned adipic acid synthesis lab activity developed by Doxsee and Hutchison is based on the Noyori protocol, which uses hydrogen peroxide as the stoichiometric oxidant and a tungsten complex as a catalyst. In addition to discussing all of the oxidizing agents mentioned previously, this lab allows discussion of some of the social and economic impacts of science.

Hydrogen peroxide is, by many measures, an ideal oxidizing agent, especially since the by-product of the oxidation reaction is water. However, there are several drawbacks to its large-scale use. First, the decomposition of hydrogen peroxide is explosive at high concentrations and temperatures. Consequently, its shipping must be in relatively dilute solutions and in refrigerated vehicles. These requirements increase solvent waste in reactions using hydrogen peroxide and result in higher energy costs. Second, although water is a benign by-product, standards for purifying water are much higher than for other chemicals since it will reenter aquifers and other stores. Furthermore, the high boiling point of water makes it costly to purify. Thus, hydrogen peroxide has not been able to supplant one of the most widely used industrial oxidants, nitric acid, primarily due to economic reasons.

Even with its many recognized drawbacks, nitric acid continues to be one of the most industrially important chemicals. It is primarily used to make fertilizers and special chemicals, including adipic acid and explosives. Very economical and convenient methods of synthesizing nitric acid were developed by the early twentieth century due to its long-standing strategic influence. In the late 1800s and in the early 1900s, nitric acid was mainly produced through treatment of Chilean saltpeter with sulfuric acid. Supplies were cut off to the Germans during World War I, which was the impetus for developing an economical version of the Ostwald process, which had been developed in 1901. In the Ostwald process, ammonia undergoes catalytic combustion, and the resulting oxide is treated with water. World War I spurred the discovery of efficient catalysts for the Ostwald process, which, combined with the Haber process, created a potent machine for the production of nitric acid. Some historians suggest that these scientific advances prolonged German engagement in World War I, costing countless lives on both sides.

Although this may case a negative light on scientific progress, it opens a dialogue with students about responsible conduct in science. Furthermore, industries that rely on nitric acid usage are easy targets for students and for their teachers. However, this discussion also provides an occasion to remind students that, we, the consumers, are part of the problem as well since we want cheaply available goods. Additionally, large pension funds, among other financial entities, depend on these corporations to turn the highest profits possible. Again, the moral and ethical ambiguity of most "real-world" situations is a message that this case study provides to students.

Catalysts

There are several important issues regarding catalysis that may be addressed during discussions of greener approaches. Through a variety of interactions, we have found that the traditional definitions of catalysts as substances that speed up reactions without being used up tend to be confusing for students. For example, during a study in which graduate students were asked to propose mechanisms to organic reactions, students often did not know how to incorporate catalysts into their mechanisms because "without being used up" translated to "without being used" in their minds (G. Bhattacharyya and G. M. Bodner, unpublished data). To help address this confusion, I suggest that catalysts are not used in the *net* reaction, which means that they are reactants in one or more elementary steps and products in others.

We have also found that students are unable to grasp the experimental implications in the definition of catalysts. A main one is that sub-stoichiometric quantities of catalysts can be used since they are not used up in the net reaction. This attribute makes them inherently greener when compared with experimental protocols that call for sto-chiometric amounts of reagents. In addition to reduction of quantity, because catalysts are supposed to speed up the rate of a reaction, one may be able to perform reactions at lower temperatures, which represent lower energy inputs. The examination of catalysts can also include the ideas of homogeneity and heterogeneity and the advantages/disadvantages of each, especially from the standpoint of separation and recovery of the catalyst from the reaction mixture. Interestingly, one of the most common catalysts in organic reactions, hydronium ion, is rarely recovered.

In addition to the experimental considerations, I also use this opportunity to discuss the differences between biological and nonbiological catalysts. Both reduce the activation energy of the rate-determining step of the process. However, they do so in markedly different ways, as illustrated in Fig. 5.4. Typical nonbiological catalysts reduce the activation energy by converting the starting material into a higher-energy intermediate. Lewis acid or protic acid catalysts, for example, tend to increase the electrophilicity of a compound by increasing the cationic character of one, or more, atoms. Though effective, these catalysts often increase the amount of undesired products since many more reaction pathways are energetically accessible to the higher-energy intermediate. In contrast, biological catalysts, typically enzymes, reduce the energy of the transition state of the rate-determining step through a variety of noncovalent interactions. Thus, enzymes reduce the energy

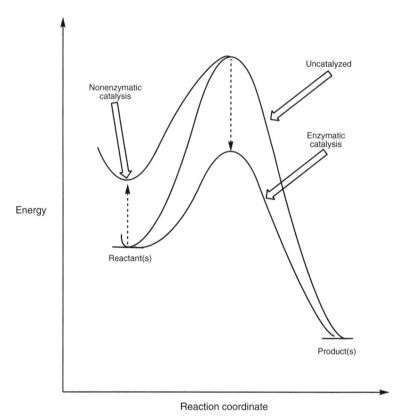

FIGURE 5.4 A schematic showing the qualitative difference between enzymatic and nonenzymatic catalysis. In enzymatic catalysis, the transition state of a specific pathway is stabilized. Typical nonenzymatic catalysis, such as protic or Lewis acid, results in increasing the energy of the starting material, making several pathways energetically accessible. This difference is the source of the selectivity found in enzymatic catalysis as well as the presence of by-products in nonenzymatic catalysis.

barrier of a specific pathway, leaving the others alone. This selectivity comes with a cost. Enzymes tend to be highly specific for certain substrates, while nonbiological catalysts tend to be more generally applicable.

PRODUCTS

This is the part of a reaction to which we tend to pay the least attention when considering greenness. Consider a scenario in which someone decides on a computer upgrade. What happens to the computer that is replaced? As another example, consider excreted, unmetabolized pharmaceuticals. Most of the studies regarding the impacts of these substances have been conducted in the context of antibiotics. Kumar et al. (2005), for example, found that antibiotics in animal feed ended up in manure, which was subsequently used as fertilizer for crops. The researchers found that plants raised on this manure had a significantly higher concentration of antibiotics as compared with those that were not exposed to the manure. Likewise, urinary excretion of antibiotics results in contamination of water sources, especially those that receive large quantities of municipal wastewater (Schiermeier 2003).

Of the several ways substances may degrade, two tend to be greener: photodegradation and biodegradation. What if we created end-market products that were easily bio- or photodegradable? Exposure to sunlight or a (micro)organism would help break down a substance without other energy inputs, as in incineration. The ideal situation would occur if the original substance were degraded into benign, nontoxic materials that could be used for future manufacturing. At the moment, leaving used materials in landfills seems to be our primary method of dealing with trash. The limitations of landfills became national news by the summer of 1987 when the "garbage barge" *Mobro* was stranded at sea after it left Islip, Long Island with over 3000 tons of New York City's trash (Gutis 1987). Although the goal was to transport the waste to an area with greater landfill space, understandably, no community would accept the vessel's load. Thus, what began as a short trip became the fodder for comedians and late-night talk show hosts.

Finding alternative, greener products represents the second main intellectual challenge posed by green chemistry. This goal challenges chemists to create substances that have the same beneficial properties as the existing option but has other characteristics that allow for its environmentally safe decomposition. There are some applications that require a specific substance. Think of some of the most effective pharmaceuticals that are on the market—there really are no substitutes. For most applications, however, there are several alternatives that have the same function. Plastics in the polymer industry are one such example. Ultimately, it will require chemists to deepen their understanding of how structure affects function. Again, all of organic chemistry will be the beneficiary, not just green organic chemistry.

SOLVENTS

Solvents are, perhaps, the greatest concern when developing greener processes. Solvents are necessary since they provide a medium for which reactions and separations/purifications can happen at an appropriate time scale. However, there is typically far more solvent than solute present in a solution, and we are typically interested in the solute rather than the solvent. Since they are generally present in large excess when compared to the reactants, reagents, and products, solvents constitute the bulk of the waste disposal.

Organic solvents can be used to cover several fundamental physical properties of chemical substances and are an ideal vehicle for discussing human toxicology. All of these ideas stem from the intended roles of solvents. Solvents should

1. dissolve all the reactants,
2. not react with any of the reactants or the products (unless it is one of the reactants), and
3. be easily removable once the process is completed.

One of the more intriguing aspects of this topic is relating molecular properties to these characteristics. Because solvents need to dissolve the reactants, which tend to be organic compounds, they have to dissolve compounds with primarily covalent bonds, many of which have low or no polarity. Therefore, highly polar substances tend not to be the most suitable solvents. But what this also means is that as polarity decreases, so do the intermolecular forces between the solvent particles. Organic solvents, therefore, tend to have low molecular weight and are lipophilic.

The second point is that solvents, by design, should be relatively inert substances. This means that they are not easily broken down by the body or in nature. Lastly, the criterion of easy removal means that most organic solvents are volatile organic compounds (VOCs). Thus, they have low vapor pressures and are easily inhaled.

In a fascinating chain of events, it is precisely these properties that also make organic solvents inherently toxic to humans. Upon exposure, there are primarily four factors that determine the substance's toxicity (Katzung 1995):

1. absorption,
2. distribution,
3. biotransformation (metabolism), and
4. elimination (excretion).

A factor common to each of these processes is that substances must past through one or more cellular membranes. Small, lipophilic molecules are the substances that pass most easily through such membranes. The key connection here is that the chemical properties that are desirable in solvents require them to be composed of small, lipophilic molecules. Thus, solvents are some of the most easily absorbed and distributed in the body. However, the most easily excreted substances are those that are water soluble. Thus, the solvents' relative inertness results in storage in the body rather than in biotransformation, which in turn prevents elimination from the body. Prolonged exposure to solvents, therefore, can result in the accumulation of a toxic concentration of that substance. This example is one of many in science in which properties that are desirable for one purpose can be detrimental for others.

Of the many greener alternative solvents that Doxsee and Hutchison (2004) suggest, water is a particularly interesting one. As previously discussed, water may not be as ideal a solvent as it appears due to high standards for release into waste streams and the difficulty in its purification. Reactions using water, on the other hand, provide yet more opportunities to emphasize the role of physical properties in chemical transformations. Ronald Breslow discovered that the low solubility of organic substances in water can actually catalyze reactions through the hydrophobic effect (Rideout and Breslow 1980; Breslow 1991). This phenomenon, known as the Breslow effect, can be a powerful vehicle for explaining the hydrophobic effect, which is one of the most important yet misunderstood ideas for future biological scientists.

OVERALL

Having covered the major components of a chemical reaction, with the exception of energy considerations, the last topic is doing an overall evaluation of a process. There are two straightforward parameters that may be calculated for a reaction that give some indication of efficiency. One is the percent yield of the purified product. The second is atom economy, which was developed by Barry Trost (1991) of Stanford University. Atom economy can be used to determine the fraction of the molecular weight of the reactant(s) incorporated into the product(s). Separate atom economies are calculated for each product of a reaction.

$$\text{Atom economy} = \frac{\text{Molecular weight of product}}{\text{Molecular weight of reactants}} \times 100$$

Higher atom economies are desirable since they imply less wastage. The discussion of atom economy can be highly illuminating for students because it can give them a very different perspective of organic chemistry reactions. Aliphatic nucleophilic substitution (S_N1/S_N2) and elimination (E_1/E_2) are two of the more emphasized reaction classes in the standard first-semester curriculum. Students often, therefore, assume these to be some of the more important and desirable reactions in organic chemistry. From an atom economy standpoint, however, elimination reactions are the most wasteful reaction type, followed by substitution. Addition reactions, in contrast, tend to have very favorable atom economies. Thus, atom economy can be used to reorient students as to preferred reaction classes in (organic) chemistry.

Of course, calculating parameters is the first step in making an overall determination of "greenness." To reiterate a theme of this chapter, evaluating a process as a whole requires students to balance several factors—including, but not limited to, toxicity of reagents and by-products, amount of waste produced, energy requirements (reaction time and temperature), yield, and atom economy—since no process is likely to be favorable in *all* aspects (Doxsee and Hutchison 2004). For example, a reaction that has a high atom economy but low yield is probably not the greenest alternative available.

One of the most powerful ways to drive home all of these points is by keeping track of waste. This can be done by several methods. Doxsee and Hutchison (2004) suggest performing an economic analysis of the entire process. For an economic analysis, students calculate all of the costs of reagents and solvents per gram of product obtained. This figure can be determined with or without the cost of waste disposal. Students are amazed to find that factoring in waste can often double the cost of production.

In addition to these techniques, there are several other sophisticated methods of analyzing the greenness of a process. For example, one can perform a life cycle assessment (LCA) of a product (Curran 2006). The LCA is used to analyze the impact of a product from the input of getting raw materials through the final disposal of the end product. Curran labels this process as a "cradle-to-grave" evaluation (p. 1). A second alternative is the environmental factor, or E-factor (Sheldon 1992, 2007). The E-factor is a measure of the waste produced per gram of product produced. It is interesting to note that due to its long tradition of chemical research, the petroleum industry has one of the lowest

E-factors, of about 0.5–1.0—that is, 0.5–1.0 g of waste produced per gram of product—and the pharmaceutical industry has E-factors as high as 1000!

CONCLUSIONS

As instructors, we strive to use "real-world" examples to help create a sense of relevance in students. However, like knowledge, the perception of reality and relevance are complex personal constructions tied to many internal factors including motivation and prior experiences. Therefore, "real-world" examples may not be "real" to everyone, and as instructors, we cannot *make* things relevant for students.

Green chemistry provides a platform through which instructors can connect with their students. As Fig. 5.1 shows, there is such a diversity of fields that relate to discussions in green chemistry; it increases the probability that students will find something meaningful in their chemistry classes since the context created by green chemistry is "real," not an amusing "aside." Thus, we hope that the "real" will be relevant.

ACKNOWLEDGMENTS

First and foremost, I would like to thank all of the students and teaching assistants of Chemistry 337 and 338 at the University of Oregon for teaching me most of what I know about green chemistry. I would also like to thank Drs. Ken Doxsee, Jim Hutchison, and Julie Haack for their help with learning green chemistry. Finally, I would like to thank Drs. Peter Wuts of Pfizer Pharmaceuticals and Dr. Doug E. Frantz of the University of Texas Southwestern Medical Center for their help with some of the industrial examples.

REFERENCES

American Chemical Society (ACS) (2008) Facts and figures: Netting the chemical industry's numbers. *Chemical and Engineering News 86*(27), 35–74.

Anastas, P. and J. Warner (1998) *Green Chemistry: Theory and Practice.* New York: Oxford University Press.

Bhattacharyya, G. and G. Bodner (2005) "It gets me to the product": How students propose organic mechanisms. *Journal of Chemical Education 82*, 1402–1407.

Bowden, K., I. M. Heilbron, E. R. H. Jones, and B. C. L. Weedon (1946) Researches on acetylenic compounds. Part I. The preparation of acetylenic ketones by oxidation of acetylenic carbinols and glycols. *Journal of the Chemical Society (London)*, 39–45.

Bowen, C. (1990) Representational systems used by graduate students while problem solving in organic chemistry. *Journal of Research in Science Teaching 27*, 351–370.

Bowen, C. and G. Bodner (1991) Problem-solving processes used by students in organic synthesis. *International Journal of Science Education 13*, 143–158.

Breslow, R. (1991) Hydrophobic effects on simple organic reactions in water. *Accounts of Chemical Research 24*, 159–164.

Chotani, G., T. Dodge, A. Hsu, M. Kumar, R. LaDuca, D. Trimbur, W. Weyler, and K. Sanford (2000) The commercial production of chemicals using pathway engineering. *Biochimica et Biophysica Acta 1543*, 434–455.

Collins, J. C., W. W. Hess, and F. J. Frank (1968) Dipyridine-chromium (VI) oxide oxidation of alcohols in dichloromethane. *Tetrahedron Letters 9*, 3363–3366.

Corey, E. J. and W. Suggs (1975) Pyridinium chlorochromate. An efficient reagent for oxidation of primary and secondary alcohols to carbonyl compounds. *Tetrahedron Letters 16*, 2647–2650.

Cornforth, R. H., J. W. Cornforth, and G. Popjak (1962) Preparation of R- and S-mevalonolactones. *Tetrahedron 18*, 1351–1354.

Curran, M. (2006) Life-cycle assessment: Principles and practice. http://www. epa.gov/nrmrl/lcaccess/lca101.html (accessed December 17, 2008).

Dess, D. B. and J. C. Martin (1983) Readily accessible 12-I-5 oxidant for the conversion of primary and secondary alcohols to aldehydes and ketones. *Journal of Organic Chemistry 48*, 4155–4156.

Doxsee, K. and J. Hutchison (2004) *Green Organic Chemistry: Strategies, Tools, and Laboratory Experiments*. Pacific Grove, CA: Brooks/Cole.

Griffith, W., S. Ley, G. Whitcombe, and A. White (1987) Preparation and use of tetra-n-butylammonium per-ruthenate (TBAP reagent) and tetra-n-propylammonium per-ruthenate (TpAP reagent) as new catalytic oxidants for alcohols. *Chemical Communications*, 1625–1627.

Gutis, P. (1987) Trash barge to end trip in Brooklyn. *New York Times*, July, 11, 1987. http://query.nytimes.com/gst/fullpage.html?res=9B0DE3DC1E3 DF932A25754C0A961948260 (accessed December 20, 2008).

Holton, R. A., H.-B. Kim, C. Somoza, F. Liang, R. J. Biediger, P. D. Boatman, M. Shindo, C. C. Smith, S. Kim, H. Nadizadeh, Y. Suzuki, C. L. Tao, P. Vu, S. H. Tang, P. S. Zhang, K. K. Murthi, L. N. Gentile, and J. H. Liu (1994) First total synthesis of taxol. 2. Completion of the C and D rings. *Journal of the American Chemical Society 116*, 1599–1600.

Katzung, B. (1995) *Basic and Clinical Pharmacology*, 6th ed. Norwalk, CT: Appleton and Lange.

Kuhn, R. and R. Torres (2002) Cre/loxP recombination and gene targeting. *Methods in Molecular Biology 180*, 175–204.

Kumar, K., S. Gupta, S. Baidoo, Y. Chander, and C. Rosen (2005) Antibiotic uptake by plants from soil fertilized with animal manure. *Journal of Environmental Quality 34*, 2082–2085.

Omura, K. and D. Swern (1978) Oxidation of alcohols by "activated" dimethylsulfoxide. A preparative, steric and mechanistic study. *Tetrahedron 34*, 1651–1660.

Ophardt, C. (2003) Virtual Chembook. http://www.elmhurst.edu/~chm/vchembook/index.html (accessed November 16, 2004).

Perry, W. G. Jr. (1970) *Forms of Intellectual and Ethical Development in the College Years: A Scheme*. New York: Holt, Rinehart and Winston.

Rideout, D. and R. Breslow (1980) Hydrophobic acceleration of Diels-Alder reactions. *Journal of the American Chemical Society 102*, 7816–7817.

Sato, K., M. Aoki, and R. Noyori (1998) A "green" route to adipic acid: Direct oxidation of cyclohexenes with 30 percent hydrogen peroxide. *Science 281*, 1646–1647.

Schiermeier, Q. (2003) Studies assess risks of drugs in water cycle. *Nature 424*, 5.

Sheldon, R. (1992) Organic synthesis – Past, present, and future. *Chemistry and Industry*, 903–907.

Sheldon, R. (2007) The E-factor: Fifteen years on. *Green Chemistry 9*, 1273–1283.

Tokay, B. (2005) Biomass chemicals. *Ullmann's Encyclopedia of Industrial Chemistry*. http://mrw.interscience.wiley.com/emrw/9783527306732/ueic/article/a04_099/current/pdf (accessed October 13, 2008).

Trost, B. (1991) The atom economy – A search for synthetic efficiency. *Science 254*, 1471–1477.

Wei, J. (2007) *Product Engineering: Molecular Structure and Properties*. New York: Oxford University Press.

CHAPTER 6

Student-Centered, Active Learning Pedagogies in Chemistry Education

GAIL MARSHALL

University of West Georgia, Carrollton, GA

INTRODUCTION

Students enroll in college chemistry classes for a variety of reasons. Some are already considering a career in science, technology, engineering, or math (STEM), but others are there because of science requirements for other areas of study. These students also come from a wide variety of high school experiences ranging from no chemistry to those who had one chemistry course, possibly with little laboratory work, to those who had significant laboratory work in an advanced course. The college experience must provide opportunities for all of these students to gain an understanding of basic chemistry concepts in a meaningful environment. Regardless of their college plans, these students also have a need for an understanding of chemistry as it applies to their roles as citizens who will be decision makers in their communities and who will need to use concepts from chemistry to better confront and influence the world in which they live.

Are chemistry classes providing these opportunities in ways that are relevant and challenging and are students able to work in a learner-sensitive environment? Although there are currently many examples in the literature of more student-oriented approaches to the teaching of college chemistry, there are still many classes where the common scene is one in which students are listening to a lecture and taking notes, possibly watching a teacher demonstration or in some cases performing laboratory work. In many of these situations, all of the students in the room are engaged in more or less the same activities and procedures, with similar achievement expectations, and most are in a relatively passive mode as they respond to the designated laboratory procedures and/or directions of the instructor. There may be some effective teaching and learning taking place, and the instructional strategies used likely have some value, but are the students also having opportunities to learn chemistry in ways that are meaningful and applicable to real life? Are the students learning to confront situations in their real world? Are the students finding intriguing possibilities for their future careers? Are they excited about what they are doing—so excited that they cannot wait to share their experiences with others? Or are they processing information for the moment or for "the test," soon to forget because they were never really engaged in meaningful learning opportunities? For the work of the students to be of utmost long-range value, the learning environment must be stimulating, providing relevant background information as well as learning opportunities for various types of students to use their minds in productive thinking as their experiences engage them in questioning, applying, analyzing, creating, and/or evaluating.

To make "chemistry relevant in a learner-sensitive environment" requires extensive planning and preparation on the part of instructors that includes not only in-depth knowledge of the chemistry concepts but also understanding of the ways and attitudes of scientists, how to identify and address differences in learners, and, most importantly, how to step out of the role of "information giver" and allow the students to truly be science investigators and learners.

STUDENT CHARACTERISTICS AND NEEDS

Typically, introductory college chemistry classes are made up of a mix of students who have selected chemistry as one of their science courses. The backgrounds, interests, and ability levels of these students can vary widely. There are even differences among those students who have more focused career goals related to chemistry. Teachers are challenged with the creation of a productive learning environment that will help to make chemistry understandable and relevant regardless of the intentions, interest levels, and/or abilities of the students.

In college classes, plans are developed and opportunities and resources selected with the assumption that students are at the same level of mental development and that all are thoroughly capable of abstract thought. At beginning college level, many students are entering into this stage of mental development, but there are significant numbers of students who are not. Some are still performing largely at the concrete level and still require experiences that provide more concrete evidence of the concepts they are studying (Williams and Cavallo 1995). Herron (1996) reported that only 50% of his college nonscience majors were thinking on the abstract or formal level.

Since chemistry deals with properties and reactions of substances, the basis for which is on a level that is not directly observable, scientists use the observable changes and products to derive theoretical models for what is happening at the unobservable level. In the most basic chemistry courses, there is often emphasis on the use of structural and mathematical modeling to represent current theories about such things as atomic structure, chemical bonding, or behavior of gases. These are important concepts, but understanding these only through theoretical models constructed by others and represented in texts or class notes can be difficult for some entering college students. These students can be helped by the thoughtful, perceptive instructor who is aware of their developmental needs and who provides opportunities for these students to also have more concrete experiences that can help them build the connections that they require.

Another aspect of the thought processes of students entering introductory chemistry courses deals with the ability to think effectively on different cognitive levels. The ability to think on these levels is not restricted to a particular stage of mental development. These levels are generally described as including factual recall and comprehension, application, analysis, synthesis, and evaluation, with the latter three or four often being described as higher levels of thinking. These are learned thinking patterns that are evident even in very young children that can be improved with experience and are not restricted to the realm of abstract thought. Students must be provided learning opportunities that will engage them in situations requiring the use of the higher levels of thinking. Only then will they grow and develop in their ability to see relevance in what they are learning, to become analytical problem solvers, to use what they are learning to add to the world of human knowledge and understanding, and to become astute consumers of information and ideas. Lectures, notes, memorization of steps and processes, and step-by-step laboratory activities provide few stimuli to promote much more than knowledge and comprehension levels of thinking. These certainly have some value in some learning situations but should not be allowed to exclude other rich opportunities for students to develop the higher levels of thinking.

THE NEED FOR MORE EFFECTIVE LEARNING ENVIRONMENTS

In an examination of how college chemistry classes are often structured, it is interesting to note some parallels in the data from studies of high school teachers' approaches to the teaching of chemistry and the potential for a cyclic effect between high school and college levels. Horizon Research, Inc. (2002) conducted a study of 5728 secondary science and mathematics teachers (74% response rate) across the United States. Among the various items included in the survey were questions about how chemistry is taught. The data indicate that the predominant forms of instructional strategies in high school chemistry courses are lecture/ discussion and practices using problems from the text and/or worksheets, approaches that are very similar to those of many college-level chemistry classes. High school teachers often explain that they are teaching to "prepare students for college," where listening to lectures, taking notes, and working mathematical problems will be important. In other words, many high school teachers are modeling their teaching after the ways in which they learned chemistry (Burke and Walton 2002). In the midst of today's reform efforts that are designed to attract more students to STEM careers, teachers at both high school and college levels are being asked to teach with different materials and in different ways than they likely were taught. The current courses, therefore, must be restructured to break the cycle of didactic, teacher-focused classes to those that are more appropriately structured to invite students as active learners. Those who are taking chemistry with the intent to teach must certainly be exposed to a newer model, one that provides them with a better understanding of chemistry concepts and how chemistry is relevant in all our lives so that they will help to break the mold of how high school students are learning chemistry. Teachers at both college and secondary levels must be ever diligent to continue to critique and to modify course structures as necessary to best prepare students for roles as decision-making citizens as well as for future careers.

The concerns of colleges and universities about the efficacy of large group lectures and verification laboratory exercises, as well as the lack of opportunities for students to experience the methods and attitudes associated with science research, have prompted the development of many alternative approaches to the teaching of college-level chemistry.

Readings in the scientific literature indicate a wide array of ideas that have been tried, dating back well into the mid-1900s, and there are likely many more that people tried but never submitted for publication. In more recent years, there has been increasing concern expressed about (1) the inability to deliver to students the ever-growing volumes of science information; (2) the lack of emphasis on the research process and the skills involved with defining a problem, gathering, analyzing, and finding meaning in data; (3) the absence of opportunities for students to experience the learning that comes from communications with others; and (4) the overall realization that fewer and fewer numbers of students are attracted to academic majors in the STEM subjects. The use of more active involvement on the part of the learner received support from the *Boyer Commission Report* "Re-Inventing Undergraduate Education: A Blueprint for America's Research Universities" (Kenney et al. 1998) and from the work done to revamp courses in a number of universities and college throughout the United States and other countries.

CONSTRUCTIVISM AND INQUIRY IN THE LEARNING ENVIRONMENT

The literature is rich with examples of approaches to college-level courses that have been designed to involve the students in more active roles as learners. There are many different versions, but among the more common are those labeled with terms such as problem based, project based, case based, or process oriented, but all of these involve the instructor more as a guide or facilitator with students who are engaged as investigative problem solvers. Each of these is based on learning approaches stemming from the ideas of constructivism and inquiry.

Constructivism arose from the work of several people, including Giambattista Vico in the early 1700s, contemporaries Piaget and Vygotsky in the early to late 1900s, and more recently von Glasersfeld in the late 1900s to the present. Although each of these proponents has a somewhat different focus, the basic idea of constructivism describes the learner's building of his or her own knowledge schema. Vygotsky, in particular, added in the concept of "social constructivism," indicating that the gaining of knowledge and understanding is enhanced through social interactions with others who are engaged in similar endeavors. Since all individuals have different backgrounds and experiences, the ways in which each individual must build onto that framework will differ. In our educational institutions, the notion of simultaneously lecturing to large groups of differing individuals or expecting all to perform more or less individually and at a similar level on common assessments is not in keeping with this model of how people learn and are motivated to learn.

The closely allied and supportive concept of "inquiry-based learning" provides an intellectual platform from which the individual may operate during the constructivist learning process. "Inquiry" comes from the same root word as the verb "to inquire," or the adjective "inquisitive," and simply means a process of questioning and looking for ways of finding information. This is something that students have been doing all of their lives but which is often stifled within classrooms at all levels. If the college instructors subscribe to the constructivist learning theory, then the classroom environment provided must be one that is sensitive to the way that learners learn and is fashioned to provide the stimuli that encourage students to engage in inquiry. In other words, the college chemistry experience must be one in which students are prompted to think and to inquire and one in which communication and the promulgation of ideas are prevalent. Students can be prompted to engage in inquiry in a number of different ways, all of which can serve different instructional purposes.

The problem- or project-based approach, along with inquiry, provide opportunities for individualization and have the most potential for use in the creation of relevant, learner-sensitive classroom environments. The lecture/demonstration, although more instructor centered, also has merit and should not be entirely excluded. It is especially appropriate for certain situations in which safety, types of materials, or time might preclude direct student investigations. Even during a demonstration, however, effective planning can indirectly engage all students and can involve all students in higher levels of thinking and in some levels of inquiry.

Bruck et al. (2008) discuss different types of inquiry, each of which should be considered for appropriate application in the chemistry classroom. These include confirmation, structured, guided, open, and authentic inquiry. The authors have developed the following rubric, which the authors describe as "useful for the undergraduate laboratory" as a guide to the levels of inquiry incorporated in the various activities and assignments. Since there is some indication of sequential development in the ability to use the various levels, great care must be taken to ensure proper alignment of student experiences within the course (Table 6.1).

Each type of inquiry-based instruction has value, and the instructor has the responsibility to know the students, the learning situation in which they are engaged, and to make appropriate judgments with regard

TABLE 6.1   A Rubric to Characterize Inquiry in the Undergraduate Laboratory (Bruck et al. 2008)

	Level 0: Confirmation	Level ½: Structured	Level 1: Guided	Level 2: Open	Level 3: Authentic
Problem/question	Provided	Provided	Provided	Provided	Not provided
Theory/ background	Provided	Provided	Provided	Provided	Not provided
Procedures/ design	Provided	Provided	Provided	Not provided	Not provided
Results analysis	Provided	Provided	Not provided	Not provided	Not provided
Results communication	Provided	Not provided	Not provided	Not provided	Not provided
Conclusions	Provided	Not provided	Not provided	Not provided	Not provided
	More structure ——————————————— Less structure				

to the type selected. *Confirmation inquiry* is probably the form most commonly used in many chemistry courses as students do prescribed laboratories that they simply follow step by step. Although confirmation inquiry can be important for initial explanations and practice of standard identification tests and safety procedures, this alone is insufficient. Students must be engaged in more than simply following directions and confirming what is already known. *Structured inquiry* steps up the challenges by allowing the students to engage in the prescribed procedures with opportunities to gather and to interpret data. *Guided inquiry* provides more opportunities for student decisions and is more appropriate when students have gained experience with laboratory procedures and skills and can, with supervision, use their own methods for collecting, organizing, and interpreting data. *Open inquiry* means that students are, within guidelines, able to select and/or to design their own ways to solve a problem, and *authentic research* describes the work of a student who is developing into an independent investigator.

The ability for and interest in engagement in "authentic inquiry" is more likely to develop in students who are having sequential experiences in the various courses that help them build the concepts and confidence that are necessary to perform at this level of independence. In view of the desire to attract more students to careers in the STEM areas, it is important to provide experiences, especially in the introductory courses, that gain and maintain student understanding and interest. It is important, therefore, to consider how these courses are taught, to look beyond the traditional, long-standing ideas about how to teach and what is expected of students to seek other more effective strategies. Based on abilities and interests, students will progress to different levels, but all can grow in their abilities in this area of mental endeavor. The instructor's role involves the planning and orchestration of opportunities that are appropriate for and available to students who are performing at different levels and who have different interests. Teaching this way is not an easy task!

PROBLEM-BASED LEARNING (PBL)

To simplify the terminology in the ensuing discussion of how chemistry courses are being made more learner sensitive and relevant, the term PBL will be used, although, as indicated earlier, there are other terms that convey a similar approach. Following this discussion, some other related approaches will be discussed in more detail. PBL can be found to describe a number of varying approaches, but all engage students in the solving of hypothetical problems in an environment that supports students' abilities to build upon what they know and to move forward from there. In other words, in the PBL class, the students are engaged in defining what they know and do not know, and then planning how to proceed, rather than being provided information and steps to follow, often attempting to move on while leaving gaps in their understanding. The variations in PBL course arrangements include not only the subjects of the problems but also the extent of the problem setting provided to the students, at what point in the course the problem is given, how the problem is introduced to the students, the materials and equipment provided, and the amount of the course devoted to this type of problem solution.

The common threads in PBL are, however, always some type of problem situation and the use of small groups of students to work on the problem. However, the presence of small groups of students working together does not necessarily indicate a PBL situation. The "test" to identify a true PBL situation is the presence of the following characteristics as outlined on the website of the University of Cincinnati:

Is the problem given to students before a lecture?

Do students work collaboratively in small groups to solve complex problems?

Are the complex problems engaging and could they occur in the "real world"?

Does student learning occur as a result of solving the problem?

Does the educator serve as a guide or consultant rather than the purveyor of information?

Does the problem require students to search for multiple sources of information?

Do the students learn new social skills in communication, cooperation, collaboration, leadership, and problem solving?

Students will tend to work most effectively and with the most diligence if the problems are of interest and are perceived by the students to have value for them. Students sometimes wonder and ask, "Why do I need to know this?" In many cases, there are multiple possible reasons why the information might be useful then or at some time in the future, but if the student cannot perceive a relationship of their experience to the real world or cannot see a definite connection to their own goals and interests, then their motivation is likely to be low. With all there is for them to learn and with limited time for them to grow as productive, contributing citizens, it is in their best interest and the interest of society at large to put great effort into making their experiences meaningful and memorable.

In a PBL setting, problems designed to be puzzling, intriguing, or that are easily perceived to relate to a significant real-world problem are introduced prior to providing information, employing the concept of *exploration before explanation*. In other words, students are allowed to "try it on their own" before the information and solutions are provided for them. Indeed, the design and intent for most PBL classes is a focus on open-ended problems that may have several alternative solutions. This is in stark contrast to the more common approach of providing the information, formulas, models, and other means of explanation of chemical concepts during a lecture, with the same concepts then possibly being reviewed in laboratory exercise, often with little or no application to real-world situations.

As outlined by Fosmire and Macklin (2002), the students' first task in any PBL setting is to understand the problem and to gain "ownership" as they, with the help of the instructor, determine what the problem entails. Students then identify what they already know and what they need to know. They then search for information, analyze what they have found, and determine if new issues have been raised. Typically, the students develop a report outlining their preferred solution, but including other potential alternatives. In most cases, the results are reported to the larger class grouping with opportunities for discussion and evaluation of the arguments.

In a PBL classroom, students take on the roles of both learner and teacher. The follow-up discussions to the work of the students, the sharing of ideas, and the questioning of the claims of others or perhaps their own claims are times of great learning. As questions are raised about the claims of a particular group, or as one group defends its own claims, it will be the students filling the board with their data and calculations or possibly discussing the equation they have developed to explain the data. The wise instructor will be intently listening, for it is during this time that clues can be gained regarding what the students have understood or where hidden misconceptions may be lurking.

It is easy to see that the instructor's role in the PBL setting changes greatly from the lecture-based class role as the giver of information and the source of reiterative questions to one who is intertwined with the learning processes of the various groups of students, listening, guiding, helping, but not directing. The instructor's position in the room changes from "up front," "on the stage" to a roving, roaming guide who is now in close proximity to the students. The instructor is now more approachable, a facilitator who listens to the work of each group and the questions the students are asking. The instructor becomes more skilled in responding to student questions with other questions that cause the students to consider what they already know and how they might find out answers for themselves. The instructor becomes skilled in the insertion of new language terms at the appropriate time, so students are gaining additional vocabulary when there is a "need to know." If, in the problem-solving process, the instructor senses a need to clarify certain points or to discuss the steps in a standard analytical procedure, a small group or entire class discussion may be conducted. The PBL instructor should realize that students gain knowledge and skills most effectively through practice and reflection, not by passively absorbing from someone else. He or she must be sensitive to the students' developmental levels and should balance appropriately the concrete and abstract content of the instruction. When using abstract formulas and equations, examples with relevant numbers rather than only algebraic variables should be provided. Throughout the instructional process, the instructor must take care to provide ample concrete illustrations and demonstrations. In a PBL classroom, the laboratory investigations help students gather information and use it to explore and to gain understanding of concepts, so there should be easy access to laboratory facilities, and care must be taken to integrate necessary laboratory training to enable students in their problem-solving efforts. The instructor's role is vital, but now, rather than being the one who is preparing for and doing most of the information giving, the instructor is helping the students with the extremely vital processes of learning to think for themselves. This requires a very different type of thinking on the part of the instructor who now needs to not only have in-depth understanding of the skills and concepts but must also be able to plan to meet the various needs of a wide variety of students as they pursue their own avenues of problem solving. At times, needs may not be anticipated and "on-the-spot" problem solving along with the students may be necessary.

The topics for PBL classrooms are obviously dictated by the objectives for the course, but the nature of the problems can be based on any of a number of scenarios written to support the course objectives and are complex enough to challenge the students. The problems should be contextual problems that are not real but could be. The problems are often described as "ill structured," meaning that the problem scenario does not provide all of the details to necessarily direct the students to a particular solution. Within the context of the problem, students should have the freedom to explore, collect, and analyze the data and to suggest a solution based on their information. Others working on the same problem may have a different interpretation, fueling an ensuing debate, perhaps a "town hall meeting," to consider the merits of the varying solutions to the problem. This type of problem is closely aligned with some topics being debated today in the United States and in other countries regarding potential solutions to environmental issues, future energy sources, access to natural resources, problems with aging populations, or the various facets of a more globalized economy. Students may sometimes originate their own problem-based investigations, especially as they become more aware of real problems and issues. This can be encouraged by providing opportunities to read and to discuss information from local newspapers, from periodicals, and/or from the Internet. The chemistry students (and others) of today are the citizens who will be dealing with these issues for a number of years into the future, so their involvement with real and important issues has potential benefits both for their learning and for the community at large.

For institutions considering the use of a PBL type of course, there are many examples in the literature for ideas and ready-made examples sufficient for entire courses. A search of university and college websites for connections to PBL can be very fruitful. At this time, some outstanding examples with excellent resources for PBL are the University of

Delaware, Duke University, the University of Illinois, the University of Cincinnati, the State University of New York, San Diego State University, Purdue, Stanford University, Western Carolina University, Samford University, and Oberlin College. Among the topics of some of the problems to be found on these sites are those dealing with home remedies, natural medicines, recently discovered fossils, spreading diseases, crimes scenes, water quality issues, chemical spills, mysterious substances, fires, explosions, climate changes, space travel, alternative fuel sources, and many others. Some of the more commonly referenced "classic" problems in the literature are "A Bad Day for Sandy Dayton: The Physics of Accident Reconstruction" (Duch 2000), "To Spray or Not to Spray: A Debate over DDT" (Dinan and Bieron 2001), "Pesticides in Drinking Water: Project-Based Learning within the Introductory Chemistry Curriculum" (O'Hara and Sanborn 1999), "Kermit to Kermette? Does the Herbicide Atrazine Feminize Male Frogs?" (Dinan 2006), "PCBs in the Last Frontier: A Case Study on the Scientific Method" (Tessmer 2005), "Case Study: Cats Have Nine Lives, but Only One Liver—The Effects of Acetaminophen" (Dewprashad 2009), and "Thalidomide Makes a Comeback: A Case Discussion Exercise That Integrates Biochemistry and Organic Chemistry" (Bennett and Comely 2001).

There are also numerous resources that can facilitate the changes that are necessary to move from larger lecture-type classes to those requiring collaboration, hands-on investigations, and the need to meet in groups of varying sizes with easy access to available resources for research. An excellent source of information can be found on the PBL section of the website for the University of Delaware. Probably the most emphatically recommended step, regardless of the information source, is the identification of course objectives. As noted by Felder et al. (2000), this delineation of objectives can have mutual benefits both for the instructor and for the students. There must be a clear delineation of the skills, attitudes, and knowledge to be demonstrated by the student upon completion of the course. In addition to the required content mastery, objectives should specify the desired emphasis on the development of performance skills necessary for finding information and for conducting scientific research. Decisions must also be made about how to evaluate student accomplishment of the objectives in the context of group problem solving. In a PBL classroom, an organizational plan for the entire course must be thoughtfully developed, for it is in this plan that determinations will be made about the extent to which PBL will be used, and this will influence the selection and sequencing of the learning activities and assignments. In the purest sense, PBL defines the entire course, but for various reasons, some institutions have utilized other models that may have several small problems throughout the course, one significant problem that is interspersed throughout the course, or possibly one culminating end-of-course problem. For PBL events, in order to model the collaboration of scientists and to maximize the social aspects of constructivist learning, plans must be made for how students can be arranged in small groups, usually of six to eight students, but with the group size, makeup, and designation of role assignments at the discretion of the instructor. Students should be given some responsibility for the operational rules for their group and possibly for rotation of responsibilities within the group. The classroom structure itself must be considered and plans developed for utilizing existing facilities if needed modifications are not possible. Changes to accommodate the new plans for teaching may also have associated financial costs. Within large classes, there may be the need to consider the assistance of undergraduate or graduate teaching assistants. Possible costs can include refurbishing classrooms, additional faculty time to develop materials, faculty attendance at training sessions, or possibly payment for consultants to assist faculty with the transition. If class sizes are altered, there may be costs for additional faculty. If new investigations are added, there may costs associated with laboratory equipment and supplies. The success of PBL programs, in the light of the changes that are generally necessitated, is greatly influenced by the degree of "buy in" by various administrators and faculty members. There is a need for a "critical mass" of the faculty to participate and to be committed to the support and enhancement of the PBL program in order to provide the momentum to keep it moving forward, to help nurture it through problem times, and to participate in the search for support to maintain and to expand the successful aspects of the program.

With regard to any new and inventive program, there is always the necessity to probe for evidence regarding the impact of the new approach. The literature has numerous reports of evaluations comparing a PBL approach to a more traditional lecture approach, and most report a more positive effect for PBL over the traditional methods. Variables in these studies are, however, difficult to control, therefore causing at least some questioning of the results. According to Lundeberg and Yadav (2006), generally, attempts to evaluate various PBL approaches are done in an effort to promote the program or are reactive to defend it against critics. While most analysts make some attempt to be quantitative, in many instances there are aspects that are difficult to evaluate. These include the type and number of the problems used and how these are distributed within the course. For example, one problem within an entire semester course, while a worthwhile experience for the students, may not have an easily detected effect on the overall performance and attitudes of the students. The enthusiasm of the instructor and statements made by the instructor may affect the attitudes and perhaps the performance of the students. Several such studies (Henderson and Mirafzal 1999; Hanson and Wolfskill 2000; Gutwill-Wise 2001) show the generally positively results in favor of at least some aspects of PBL. Positive ratings are generally associated with students' reports of satisfaction with the course content and presentation. When compared to more traditional lecture courses, attendance and enrollment in subsequent courses have been shown to increase, and increases in exam scores were observed. Several other papers describe somewhat more sophisticated analytical techniques applied to large numbers of instances where PBL was used and with generally the same positive results for PBL. Dochy et al. (2003) report on a meta-analysis of data from 43 reports. There was a definite positive effect from PBL on the skills of the students and no study reported negative effects. There was a less conclusive effect on knowledge, the overall results being strongly influenced by two studies with negative results. Strobel and van Barneveld (2008) describe a qualitative metasynthesis approach to evaluate assumptions of meta-analytical research on the effectiveness of PBL. The results indicate that PBL is more effective in promoting long-term retention of knowledge, greater skill development, and higher levels of satisfaction on the part of students and teachers. More traditional classroom approaches are effective for short-term retention. Walker and Leary (2008) and Ravitz (2008) report on an analysis of 47 data sets from PBL implementation outside the medical and allied health fields. This is significant because for many years, the predominant use, therefore the major focus of statistical analyses, was in health-related areas. The results of this study indicate for most of the analyses that students involved in PBL classes did as well as or were better than those in lecture-style classes. In subjects outside the medical field, the tendency was for PBL students to do better than students in lecture classes. As the efforts to improve teaching in chemistry (and other disciplines) has continued, the exact design of PBL and the other plans with similar names have taken on variation. The authors suggest that future assessments and evaluations be directed more toward the identification of particular forms of PBL and the supporting mechanisms and strategies to facilitate the implementation of PBL.

PROCESS-ORIENTED GUIDED INQUIRY LEARNING (POGIL)

A closely related but different approach to the improvement of science instruction is termed POGIL. This is a National Science Foundation (NSF)-developed program for the improvement of undergraduate chemistry courses that combines a guided inquiry approach and a conscious development of student process skills. The success has even spilled over into the development of a high school chemistry "POGIL" program. POGIL uses a somewhat more structured definition of "guided inquiry" PBL by encouraging teachers to follow a "learning cycle paradigm" with an approach that consciously develops particular learning skills in students. Nevertheless, the approach emphasizes the same PBL concepts of relevancy and attention to the needs and interests of individual learners. The POGIL program was first implemented for general chemistry classes at Franklin and Marshall College in 1994. Assessment of the effectiveness (Farrell et al. 1999) indicated that attrition in general chemistry dropped from 22% to 10%, and the average percentage of students earning and A or B increased from 52% to 64%. Reports of their tremendous success resulted in an NSF grant in 2003 for the POGIL Project, based at Franklin and Marshall, and launched a national program for dissemination. Subsequent evaluations indicate similar results at both small and large universities. At Harvey Mudd College, POGIL is used for the general chemistry course (Karukstis 2003). Case studies provide a student-centered problem base to help students examine the relationships between chemistry and society. In small teams, they explore environmental and economic implications of decisions in cases involving industrial chemical plants, mining operations, and the design and construction of batteries for pacemakers. Students are forced to use scientific approaches to the investigation of open-ended problems that are very much real-life problems in the world today. POGIL applied in general chemistry courses at the State University of New York at Stony Brook resulted in significantly higher attendance at recitation sessions (from 10–20% to 80–90%), most students (75–90%) reporting that the workshops are challenging and worthwhile, improved examination scores (20% more students scoring above 50% level), increased enrollments in the second year organic chemistry course (up by 15%), and anecdotal comments by the instructors regarding improvement in skills throughout the semester regarding reading and understanding the text, applying concepts to solve problems, teamwork, oral and written communications, and reflecting on learning and performance as an individual and as a team member (Spencer et al. 2003).

PEER-LED TEAM LEARNING (PLTL)

In addition to PBL and POGIL, a third related approach to the attempts to improve instruction in college chemistry courses is labeled PLTL. In this model, a student who has previously performed well in the course is a guide for teams of six to eight students in learning sciences, mathematics, and other undergraduate disciplines. The PLTL Workshop model is designed to provide an active learning experience for students, to create a leadership role for undergraduates, and to engage faculty in a creative new dimension of instruction. As described by Eberlein et al. (2008), PLTL is designed to supplement, not to supplant, lecture time. The collaboration of the workshop participants and the peer leaders provides opportunities for active student participation that is very likely not available within the lecture class. PLTL has been used successfully in many different disciplines and has the advantage of offering a mix of active learning opportunities for students and a new way to enhance the growth and leadership development of promising undergraduates in a particular discipline. For the peer leaders, the experience of working with faculty and guiding their peers through a difficult course is rewarding and unforgettable, and can have a profound effect on their individual and professional growth.

Six critical components for successful implementation of the PLTL instructional method include the following: (1) the workshop as a regular course component that all students are expected to attend; (2) the faculty teaching the course is closely involved with the workshops and the workshop leaders; (3) the workshop leaders are well trained and are closely supervised, with attention to content knowledge and teaching and learning techniques; (4) the workshop materials are challenging and encourage collaborative problem solving; (5) the organizational arrangements are optimized to promote learning; and (6) there is appropriate institutional support for innovative teaching.

While the intent of this chapter is not a detailed discussion of the PLTL approach, it is important to recognize that this can be an integral part of any attempt to revise a chemistry curriculum to be more learner centered and more sensitive to the needs of individual learners. Even some of the PBL or POGIL classes are still organized within large group settings. The needs of individuals may sometimes be obscured among the large number of students with whom an instructor must work, even if that instructor is taking on the role of guide, moving about the room, and having generally more proximity to the students. As a part of a planned course revision, PLTL is certainly an approach that should be investigated, but which will also require considerable attention to how it is integrated with the rest of the course, how the leaders are trained, and how the instructors view and support the workshops. Much more information, as well as links to papers about PLTL, can be found at the website for PLTL at the City College of New York, the place where PLTL began.

SUMMARY

There seems to be an established, understood need to make introductory college chemistry courses much more meaningful and relevant for all types of students. Given what is known about how people learn, the variations in mental development of college-age students, what helps students retain what they are learning, and the needs that people have for social engagement while learning, the currently predominant large group, one-size-fits-all classrooms are not appropriate. Furthermore, the rapidly advancing amount of knowledge makes decisions about what to teach more and more difficult. Those who plan introductory chemistry courses should consider the potential advantages of some version of problem-based, student-centered learning designs. The chemistry classrooms of today must be designed to provide supportive environments where students' opportunities to learn can be optimized, where they can learn to work with others, and where they can grow in the ability to apply scientific reasoning and problem solving. Optimistically, the numbers who decide to study chemistry as a major will increase and they will be better prepared for subsequent courses. Hopefully, those who go on to other studies will have a good understanding of how chemistry is important in the lives of everyone.

REFERENCES

Bennett, N. and K. J. Comely (2001) Thalidomide makes a comeback: A case discussion exercise that integrates biochemistry and organic chemistry. *Journal of Chemical Education 78*, 759.

Bruck, L. B., S. L. Bretz, and M. A. Towns (2008) Research and teaching: Characterizing the level of inquiry in an undergraduate laboratory. *Journal of College Science Teaching 9*, 52–58.

Burke, B. A. and E. J. Walton (2002) Modeling effective teaching and learning in chemistry. *Journal of Chemical Education 79*, 155.

Dewprashad, B. (2009) Case study: Cats have nine lives, but only one liver— The effects of acetaminophen. *Journal of College Science Teaching 38*(7), 48–52.

Dinan, F. J. and J. F. Bieron (2001) To spray or not to spray: A debate over DDT. *Journal of College Science Teaching 31*(1), 32–36.

Dinan, F. J. (2006) Kermit to Kermette? Does the herbicide atrazine feminize male frogs? *Journal of College Science Teaching 35*(10), 38–42.

Dochy, F., M. Segers, P. Van den Bossche, and D. Gijbels (2003) Effects of problem-based learning: A meta-analysis. *Learning and Instruction 13*(5), 533–568.

Duch, B. J. (2000) A bad day for Sandy Dayton: The physics of accident reconstruction. *Journal of College Science Teaching 30*(1), 17–21.

Eberlein, T., J. Kampmeier, V. Minderhout, R. S. Moog, T. Platt, P. Varma-Nelson, and H. B. White (2008) Pedagogies of engagement in science: A comparison of PBL, POGIL, and PLTL. *Biochemistry and Molecular Biology Education 36*(4), 262–273.

Farrell, J. J., R. S. Moog, and J. N. Spencer (1999) A guided-inquiry general chemistry course. *Journal of Chemical Education 76*, 570.

Felder, R. M., D. R. Woods, J. E. Stice, and A. Rugarcia (2000) The future of engineering education: Teaching methods that work. *Chemical Engineering Education 34*(1), 26–39.

Fosmire, M. and A. Macklin (2002) Riding the active learning wave: Using problem-based learning as a catalyst for creating faculty-librarian partnerships. *Issues in Science and Technology and Librarianship 34*, 1–10.

Gutwill-Wise, J. P. (2001) The impact of active and context-based learning in introductory chemistry courses: An early evaluation of the modular approach. *Journal of Chemical Education 78*, 684.

Hanson, D. M. and T. Wolfskill (2000) Process workshops—A new model for instruction. *Journal of Chemical Education 77*, 120.

Henderson, L. L. and G. A. Mirafzal (1999) A first-class-meeting exercise for general chemistry: Introduction to chemistry through an experimental tour. *Journal of Chemical Education 76*, 1221.

Herron, J. D. (1996) *The Chemistry Classroom*. Washington, DC: American Chemical Society.

Horizon Research, Inc. (2002) *2000 National Survey of Science and Mathematics Education: Status of High School Chemistry Teaching*. Chapel Hill, NC: Horizon Research.

Karukstis, K. K. (2003) Using case studies to introduce environmental and economic considerations in general chemistry. Presented at Non-Traditional Teaching Methods: Methods Other Than Lecture and Assessment of These Methods, an online conference, March 28–May 9, 2003.

Kenney, S. S., E. Thomas, W. Katkin, M. Lemming, P. Smith, M. Glaser, and W. Gross (1998) Re-inventing Undergraduate Education: A Blueprint for America's Research Universities. *Boyer Commission Report on Educating Undergraduates in the Research University*. Stony Brook, NY: State University of New York.

Lundeberg, M. A. and A. Yadav (2006) Assessment of case study teaching: Where do we go from here? Part II. *Journal of College Science Teaching* *35*, 6–13.

O'Hara, P. B. and J. A. Sanborn (1999) Pesticides in drinking water: Project-based learning within the introductory chemistry curriculum. *Journal of Chemical Education 76*(12), 1673.

Ravitz, J. (2008) Introduction: Summarizing findings and looking ahead to a new generation of PBL research. *Interdisciplinary Journal of Problem-Based Learning 3*(1), 4–11.

Spencer, J. N., R. S. Moog, F. J. Creegan, D. M. Hanson, T. Wolfskill, A. Straumanis, and D. Bunce (2003) Process oriented guided inquiry learning. Presented at Non-Traditional Teaching Methods: Methods Other Than Lecture and Assessment of These Methods, an online conference, March 28–May 9, 2003.

Strobel, J. and A. van Barneveld (2008) When is PBL more effective? A meta-synthesis of meta-analyses comparing PBL to conventional classrooms. *Interdisciplinary Journal of Problem-Based Learning 3*(1), 44–58.

Tessmer, M. (2005) PCBs in the last frontier: A case study on the scientific method. *Journal of College Science Teaching 34*(10), 34–36.

Walker, A. and H. Leary (2008) A problem based learning meta analysis: Differences across problem types, implementation types, disciplines, and assessment levels. *Interdisciplinary Journal of Problem-Based Learning 3*(1), 12–43.

Williams, K. A. and A. M. L. Cavallo (1995) Reasoning ability, meaningful learning and students' understanding of physics concepts. *Journal of College Science Teaching 24*(5), 311–314.

CHAPTER 7

Creating a Relevant, Learner-Centered Classroom for Allied Health Chemistry

LAURA DELONG FROST

Georgia Southern University, Statesboro, GA

Many allied health courses of study have a chemistry component within their programs. Chemistry departments often view such courses as service courses and many times assign one faculty member (or a committee) to spearhead their course development. This chapter represents advice and guidance for faculty attempting to develop an entry-level allied health chemistry course for nonmajors. The chapter begins by reviewing literature that examines the challenges in developing (1) appropriate course content, (2) teaching strategies for student learning and engagement, and (3) the elements of a classroom environment conducive to learning. A successful working model for this course implementing best practices is then provided based on the author's own experiences. Course assessment after six semesters indicates enhanced student learning and higher student perception of learning following implementation.

STRATEGIES

Organizing Course Content

The chemistry requirements for allied health programs vary depending on program, and they also vary from institution to institution. Allied

Making Chemistry Relevant: Strategies for Including All Students in
A Learner-Sensitive Classroom Environment, Edited by Sharmistha Basu-Dutt
Copyright © 2010 John Wiley & Sons, Inc.

health programs tend to have fairly rigid course frameworks with little room for electives. As an example, some nursing programs require general chemistry; some require two semesters of chemistry covering aspects of general, organic, and biological chemistry (GOB); other nursing programs allow just one semester for this coverage, and still other nursing programs have removed the requirement completely.

What do students in an allied health program such as nursing actually need to know about chemistry? What should be covered? What should be left out? Chemists have been trying to determine appropriate content for over 30 years! In 1979, the American Chemical Society organized a task force on chemical education for the health professions. In the 1980s, this task force published a syllabus of major topics considered important for health professions that should be included in a one-semester course (Treblow et al. 1984) and later a two-semester course (Daly and Sarquis 1987). To summarize the recommendations of the task force, their syllabus contained over 50 topics with subtopics, limited the organic topics, and recommended that time be devoted to biochemical topics at the end of the course. It is interesting to note that even as early as 1984 the task force recommended organic topics be limited to selected functional groups and redox reactions.

In 1992, a survey of deans of schools of nursing or department heads and practicing nursing was published by Walhout and Heinschel (1992). In this survey, these two groups were asked to rank 39 topics covered in allied health chemistry courses (as determined by chemists) as to their importance. The highest ranked chemistry topic from this list was *salts and buffers* with *acids, bases, and pH* ranking third. The rest of the top ten topics were biochemical topics with organic chemistry topics receiving the lowest rankings. In a separate study conducted around the same time, Dever (1991) found that faculty members teaching in the allied health disciplines felt that chemistry educators "spend too much time on organic chemistry." Walhout and Heinschel suggest that chemistry teachers need to consider whether we are trying to cover too much in our chemistry classes. Perhaps the "less is more" policy will have the effect of building better foundations for this cohort's future courses.

More recently, at the 2004 *Biennial Conference on Chemical Education*, nursing faculty Dolter and Marshall (2004), then at the University of Iowa, described their use of focus groups to enumerate the chemistry content essential for the nursing practice. Their findings again indicated that biochemical topics were the most important to this group of students and they specifically indicated that membranes should be covered.

Many allied health programs (and more specifically nursing programs) are being forced to reduce a two-semester chemistry requirement into a one-semester requirement and are asking chemistry departments to produce a reasonable course for these students. One of the few reports on such a task was described by Henry Tracy in 1998. After an examination of the literature described, he condensed his course down to four categories: measurement, molecular structure, acid–base chemistry, and biochemistry. He allocates appropriate time to each category and ties all the material together with a molecular modeling term project.

In 2006, myself, Deal, and Humphrey describe our integrated curriculum for a one-semester GOB course where general and organic chemistry topics are discussed in the context of biochemical concepts and applications. Student interest in a course organized in this way was significantly increased using the integrated curriculum versus the former "follow the traditional textbook"[1] curriculum. This interest was shown to be instructor independent.

In summary, much of the literature surrounding allied health chemistry requirements advocates a selection of GOB topics focusing on a few key general chemistry topics like solution and acid–base chemistry as well as several biochemical topics due to their applicability and relevance.

Approaches for Student Engagement and Learning

In addition to showing interest in the course, students must engage at some level with the subject matter for learning to take place. Motivating students to learn can be particularly difficult for a droning lecturer who dims the lights and reads PowerPoint slides. The efficient delivery of material via a lecturing professor does not automatically translate into efficient learning by the student (Johnson et al. 1991a; Holme 1993; Ward and Bodner 1993; Zoller 1993; King 1994). This is especially true in the sciences where many topics are highly abstract and often students do not see the application in what they are learning. Students have real difficulties applying knowledge to solving textbook, exam, and, eventually, real-world problems. In addition, students in the lecture setting gain little experience in teamwork and associated skills needed in the workplace (Hanson and Apple 2004). As this book suggests, there are a number of viable approaches that can be used to engage students. Three commonly used student-centered, active learning pedagogies in

---

[1]That is, general chemistry followed by organic chemistry followed by biochemistry.

science education are problem-based learning (PBL), peer-led team learning (PLTL), and process-oriented guided inquiry learning (POGIL) (Eberlein et al. 2008). The POGIL approach is discussed here because it is the pedagogy used in the author's allied health chemistry course.

POGIL is a guided inquiry, non-lecture-based, cooperative learning approach rooted in constructivist learning theory. Students are encouraged to construct new knowledge through a learning cycle that allows them to explore a model, construct concepts, and apply these concepts to new situations (Piaget 1964; Karplus and Their 1967). In addition to delivering chemistry content to students, POGIL has a process component that helps students develop skills for acquiring, applying, and generating knowledge. These components to critical thinking are essential for allied health students as they move through their programs and provide diagnoses for patients. The POGIL classroom consists of groups of three to five students working on deliberately designed guided inquiry materials. These activities typically supply students with data followed by critical thinking questions that allow students to formulate their own conclusions—essentially constructing knowledge for themselves through scientific inquiry. The instructor serves as a facilitator of learning in this setting rather than an information source.

Does the POGIL approach increase student learning in chemistry? An examination of the POGIL classroom has been shown to increase student mastery of content over the traditional lecture approach (Farrell et al. 1999; G. McKnight, unpublished data) for science majors. Restructuring the classroom from traditional lecture to an actively involved student–student and student–instructor group setting such as that fostered by POGIL has also been identified as having the largest positive effect on a number of environmental factors on academic achievement, personal development, and satisfaction of college students (McKeachie et al. 1986).

Selection of an appropriate teaching approach should be an individual choice as effective instructors have a variety of styles. However, student engagement in the classroom in some form is critical to student learning.

Developing a Classroom Learning Environment

One of our roles as educators is to create a classroom environment conducive to learning. There are many ways to create an environment conducive to learning. Traditionally, this has involved simply speaking to students. In her book *Learner-Centered Teaching*, Maryellen Weimer

(2002a) provides a review of several student studies on classroom environment preferences. These studies agree that when students are in a classroom environment they prefer, they achieve more. Students do not find teacher-controlled, rule-oriented, requirement-driven classrooms ideal. Weimer pinpoints "opportunity for instructor–student interaction" as an essential component of a functional learning environment. Instructors must take action to create these opportunities.

Regarding classroom environment, Ken Bain (2004) offers this observation in his book *What the Best College Teachers Do*:

> More than anything else, the best teachers try to create a natural critical learning environment: "natural" because students encounter the skills, habits, attitudes, and information they are trying to learn embedded in tasks they find fascinating; - authentic tasks that arouse curiosity and become intrinsically interesting; "critical" because students learn to think critically, to reason from evidence, to examine the quality of their reasoning using a variety of intellectual standards, to make improvements while thinking, and to ask probing and insightful questions about the thinking of other people.

Bain notes that natural critical learning is rooted in critical thinking and active learning strategies. Active learning recognizes that students must be involved in their own learning, while natural critical learning recognizes that the action is most effective if the learner decides to do it because it satisfies a need to know—not just because he or she was told to go talk to a neighbor.

Bain further identifies five essential elements to a natural critical learning environment: (1) offer an intriguing question, (2) guidance as to the importance of the question, (3) engagement in higher-level intellectual activity (higher than listen and remember), (4) help in answering the question, and (5) development of further questions by the students. The POGIL approach described encompasses many if not all of these elements.

Setting up a classroom learning environment that deviates from the traditional classroom environment often meets with resistance in many forms. Weimer (2002b) offers the following tips for overcoming resistance to a new classroom environment:

- Discuss the rationale behind your classroom environment openly and regularly.
- Offer encouraging, positive reinforcement. Your belief in the students' ability to learn is central.

- Ask for feedback and communicate your comments to the feedback.
- Do not become part of the problem by becoming uncaring.

Appropriate structuring of the classroom environment can modify student perceptions of learning and teaching effectiveness. In the allied health classroom, often female students are in the majority, so learning styles based on gender should be considered. The literature offers the following guidelines: women tend to be socialized to be more supportive and collaborative versus authoritarian and hierarchical,[2] so structuring a classroom within this social framework using class group work can be useful for this group of students. Female students also tend to prefer organized course materials, knowledgeable instructors, and reading assignments as opposed to male students who prefer "hands-on" learning tasks and a more pragmatic approach to learning (Pettigrew and Zakrajsek 1984). Females prefer a more conceptual learning style where males prefer an applied learning style (Keri 2002).

Creating a classroom environment conducive to learning can be measured both by increased student learning of content as described above and by student perceptions of learning. Student perceptions of learning have also been shown to be a reflection of instructional effectiveness (Centra and Gaubatz 2000).

As the literature shows, the classroom environment developed by the instructor can affect student perceptions, teaching effectiveness, and learning.

SUCCESSFUL IMPLEMENTATION

As background, at Georgia Southern University (GSU), a one-semester GOB course has been offered to allied health majors since converting to the semester system in 1998. Allied health at GSU consists of mostly nursing followed by nutrition majors. Additionally, up to 30% of the students enrolled in a section of this course take the one-semester GOB course as a general education requirement.

Advice and evidence of enhanced student learning after overhauling the course content, teaching approach, and classroom environment are offered below.

[2]For a review, see Maher F. A. and M. K. T. Tetreault (1994) *The Feminist Classroom*. New York: Basic Books.

ORGANIZATION

Based on the advice from the literature and from colleagues as well as some personal trial and error, it seemed appropriate to reorganize the course content for a GOB course by de-emphasizing some topics, integrating others, and providing applications in order to create a curriculum that both provides the necessary fundamentals of chemistry and includes topics most relevant to the health science student. Karen Timberlake (2004), author of leading textbooks for this class of students, offers the following rule of thumb for topic selection: include topics essential to chemistry with high value (applicability) to student careers and reasonable learning time while omitting topics that have less impact on understanding of future concepts and less value to student careers.

Elaboration of the strategy employed for organizing course content for an allied health chemistry course follows.

De-Emphasize Some Topics

There are a number of chemistry topics that, although they must be well understood for science majors preparing to extend their knowledge further in the discipline, appear irrelevant to allied health majors because they have no application to their real-world problems. As an example, the differing shapes of atomic orbitals is a fundamental chemical concept that is (and should be) taught in any general chemistry curriculum. Allied health students, however, do not need to go on to understand valence bond theory or molecular orbital theory, and so have no need to understand this concept. Much of their understanding of chemical bonding involves only the elements carbon, oxygen, nitrogen, and hydrogen and should be presented in that context. It is more important to consider the audience's need to know than the instructor's need to present.

Several topics are suggested here that could be de-emphasized or eliminated, affording instructors time to explore biochemical topics more fully: electron configuration, quantum numbers, atomic orbitals, the mole concept, limiting reactant and stoichiometry, organic nomenclature, and organic reactions by functional group.

Integrate Topics

Wherever possible, general and organic chemistry topics should be tied to their biochemical relevance. As an example, redox reactions, which

TABLE 7.1   Chemical Principles and Their Health Applications

Chemical Principle	Health Application
Boyle's law	Breathing, "the bends"
LeChâtelier's principle, buffers, pH	Bicarbonate buffer system, acidosis, and alkalosis
Radioactivity	Radioisotopes, nuclear medicine
Osmosis	Cell crenation/rupture
Diffusion	Ion transport
Stereochemistry	Chiral drugs (historical example: thalidomide)

are very important in biological systems, can be introduced in the context of reducing sugars, disulfide bonds in cystine, and the oxidized and reduced nucleotides such as $NAD^+/NADH$.

Provide Application

Because the allied health student is interested in the application of chemical concepts on their discipline, it is useful to incorporate chemistry applied to health into the curriculum. Many times, such applications can be found in the lesser read text boxes in traditional GOB textbooks. A few examples are offered in Table 7.1, but many more exist.

As faculty, it is important to remember that these students need relevance and applicability in order to become good professionals; otherwise, chemistry becomes just another hoop that they must jump through to be successful in their program.

GROUP LEARNING IN CLASS THROUGH GUIDED INQUIRY

The POGIL approach has been used in the author's sections of the one-semester GOB course previously described since the fall of 2006. This section describes key elements in such an approach and notes the best practices employed by the author. Student learning using this guided inquiry approach is then described.

Best Practices

Elements used in a guided inquiry approach such as POGIL include the following:

- learning teams,
- in-class activities,

- metacognition and reporting, and
- individual accountability.

Learning Teams   There is ample evidence in the literature that students working in cooperative teams learn and understand more, have a more positive attitude regarding the subject matter, and feel better about themselves and about the class (Johnson et al. 1991b; Slavin 1995). The POGIL approach employs cooperative learning teams of ideally four students where each student is assigned a role that assists the group in completion of the learning activity (Hanson 2006). Rotating the groups during the semester has the advantage of allowing students to work with other students, distributes "freeloading" students, and breaks up predictable patterns that students may fall into if left in the same group. However, as Michaelsen et al. (2004) points out, in order to form a cohesive team that can function effectively, it is useful to have students remain in the same group for the entire semester. In the allied health course described here, students are often satisfied with the random groups organized on the first day of class. They become comfortable with their group members and when asked, they do not wish to rotate. As the semester progresses, these initial groupings develop into cohesive teams.

In-Class Activities   A full semester of guided inquiry activities covering a range of GOB topics was developed by the author since the integrated course content described earlier precluded the use of the current commercially available POGIL activities (Garoutte 2007). An example is shown in Fig. 7.1. The worksheet begins with a table of sample data. This is followed by a set of critical thinking questions guiding the student to first explore the data (questions 1–3), to invent their own chemical concept (questions 4 and 5), and to apply the concept invented to other data not shown in the table (question 6). This thought process parallels the constructivist learning cycle (Piaget 1964; Karplus and Thier 1967) and is integral to the design of inquiry-based activities. Other key components of an activity include textbook references so that students can review concepts developed in class with their textbook and homework problems that both reinforce the concepts from class and ask students to apply those concepts to a more challenging situation.

In a guided inquiry classroom, the instructor's role shifts from being the source of information typical in a lecture setting to that of a facilitator for learning. As students work on activities during the

What are the three temperature scales?
Can you convert between them?

Three Temperature Scales

	Kelvin (K)	Celsius (°C)	Fahrenheit (°F)
Boiling point of water	373	100°C	212
Freezing point of water	273	0°C	32
Difference between the boiling and the freezing points of water	100	100	180

Examine the data shown for the three temperature scales and the differences between the freezing and boiling points of water. Then, answer the questions.

Questions

1. How is the size of a single kelvin related to the size of a single degree Celsius? (Hint: What is the difference between the freezing and boiling points of water?)

2. Based on Question #1, provide an equation for converting from kelvin to Celsius.

    °C= _____

3. Consider the following: There are 10 dimes or 100 pennies in one dollar. A dime is worth 10 times more than a penny, or we could say that there is 1 dime per 10 pennies (1 dime/10 pennies)

    a. Similarly, which is larger: a degree Celsius or a degree Fahrenheit?

    b. By how much (provide a fraction)? This fraction is considered a conversion factor.

4. To convert from degrees Fahrenheit to degrees Celsius, you would need to subtract 32 to get to the reference point (0°C), and then use a conversion factor (determined above) so that the units cancel appropriately. Write an equation for this.

5. To convert from degrees Celsius to degrees Fahrenheit, you would need to use the appropriate conversion factor to convert to degrees Fahrenheit first, and then add 32°F units. Write an equation for this.

6. A refrigerator's temperature is 4°C. What is that in degrees Fahrenheit?

FIGURE 7.1   A sample POGIL exercise.

classroom, the instructor circulates through the classroom serving as a clarifier of material. Instructors should not directly provide answers to the critical thinking questions. At first, this may be difficult for some instructors. Group answers can be reported to the class for acceptance. This has the effect of helping to establish cohesive groups and building confidence. Some instructors choose to grade the group activities, while others choose some form of individual accountability (see below). The author prefers individual accountability in the allied health course.

Metacognition and Reporting   POGIL uses metacognition to help students self-regulate, self-assess, and reflect on their learning (Hanson 2006). Students are asked to provide written daily feedback regarding their learning: What did they understand? What didn't they understand from class today? Instructors must read student feedback prior to the next class period to address student understanding in a timely manner.

Individual Accountability   In order to ensure that students are contributing to their teams, individual accountability must also be a component of any group learning approach. This can be achieved by grading homework problems or by administering a short daily quiz. The author uses the latter, offering a short quiz (taking up no more than 10 minutes) at the beginning of the class period based on material from the previous period. This quiz is immediately reviewed, offering a teachable moment before moving on to the next topic.

ASSESSING STUDENT LEARNING

Does the use of a cooperative, guided inquiry-based approach like POGIL affect student learning? Student learning using the POGIL approach was assessed by examining course grades, final exam grades, and questions on the final by topic and by learning level. All sections (POGIL and control) were taught by the author using the integrated curriculum described above. The control classroom used a previous teaching approach where the classroom engaged students through active learning handouts that could be described as "practice" after concept introduction by a PowerPoint lecture. The control approach will be referred to as lecture-interactive (L-I).

Course Grades

Course grades can be an initial indicator of student learning but, since a course grade evaluates other aspects of the course, should not be relied on exclusively to produce an argument for increased student learning in a course. However, if all other aspects remain constant (e.g., the lab portion of this course remained the same and is included in the final course grade), an increase in student course grades *can* imply greater student learning in a course.

Figure 7.2 shows a comparison of final course grades for the first six semesters using POGIL ($N = 271$) versus a similar sample size of the earlier L-I approach ($N = 285$). This data shows that course grades overall have shifted; the number of students being successful in the

Final grade distribution

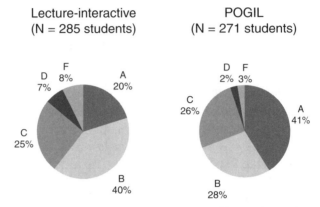

Lecture-interactive
(N = 285 students)

POGIL
(N = 271 students)

FIGURE 7.2 Final grade distributions for comparable numbers of students taught using POGIL versus the previous lecture-interactive (L-I) approaches. The L-I sections (six) spanned five semesters from Fall 2002 to Fall 2005, while the POGIL sections (seven) spanned six semesters from Fall 2006 to Summer 2008. The number of withdrawals remained constant at ~8% for both groups.

TABLE 7.2   Comparison of DFW Rates, L-I versus POGIL

	% DFW	% Withdrawal Only
Lecture-interactive (N = 304)	20	8
POGIL (N = 306)	17	8
POGIL—first semester (N = 96)	24	9
POGIL—second to sixth semesters (N = 210)	14	8

course (passing with a C or better) has increased using the POGIL approach. When this is examined in conjunction with the number of students failing or withdrawing from the course, the DFW rate (Table 7.2), an overall comparison shows that withdrawal rates remain fairly steady regardless of the approach, whereas the number of failing grades in the course decreases under POGIL following the first semester. This interesting result can be explained by increased student discomfort (students have not had or heard of any science courses being taught in this way) and less effective facilitation by the instructor during the first POGIL semester.

Final Exam Grades

Offering a common final exam using two different teaching approaches gives a more direct measure of student performance in a course. During

TABLE 7.3   Comparison of Final Exam Scores, L-I versus POGIL

	Final Exam Score 90% Similarity		Final Exam Score Increasingly Difficult in POGIL
Lecture-interactive Fall 2005 ($N = 91$)	58.3 ± 16.1	Lecture-interactive ($N = 274$)	60.4 ± 15.9
POGIL Fall 2006 ($N = 81$)	64.8 ± 14.2	All POGIL ($N = 266$)	65.1 ± 13.3

the fall of 2005 (L-I) and the fall of 2006 (POGIL), a multiple choice final exam that was 90% similar was given to the two groups. The average score on the final exam was significantly higher for the POGIL group than for the L-I group (Table 7.3). After the fall of 2006, the POGIL final exams included more challenging exam questions (determined by question learning level—see below). Despite this increase in difficulty of the final exam, when the average final exam score for *all* POGIL sections is compared to a comparable group of L-I students, the POGIL students continue to outperform the L-I students.

Questions on the Final Exam

To determine the level of difficulty of the final exams, common questions from the final exams (Fall 2005–Summer 2008) were categorized by their learning level. For this analysis, questions were placed into one of four levels as defined by Gagné and Briggs (1974). These levels are summarized below from lowest to highest:

1. Recall of information—basic memorization
2. Classification of concepts—application of information formulating a single concept or principle
3. Demonstration of rules—application of multiple concepts solving simple problems
4. Generation of higher-order rules—application of multiple rules solving complex problems

After categorizing common questions from several semesters of final exams, Fig. 7.3 shows that the earlier final exams, including the L-I exam, included more questions at the lower learning levels and did not include questions at the highest level of learning (higher-order rules). Despite this increase in difficulty, POGIL students scored significantly higher on their final exam than the L-I students.

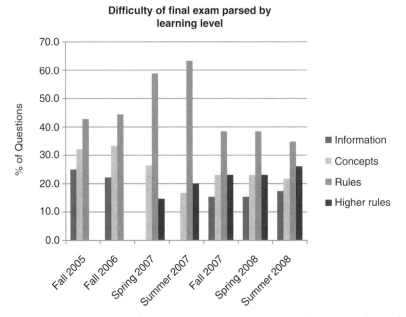

FIGURE 7.3   The level of difficulty of the questions on the final exam was determined by categorizing 24–32 common questions on the final exam by learning level as described by Gagné and Briggs (1974). The exams given to the POGIL sections (Fall 2006–Summer 2008) became more difficult.

To examine actual student learning of chemical topics, the same set of common questions from the final exam was categorized by chemical topic. The percentage of students getting a complete correct answer to the questions was determined for 24–32 final exam questions covering up to 16 topics.[3] The difference in the percentage of students getting a topic correct was determined for the two student groups (POGIL vs. L-I). Figure 7.4 shows the students in the POGIL sections got more topics correct more often on the final exam, an indication that they learned more chemistry.

In summary, does the POGIL approach increase student learning in chemistry for nonmajors? When allied health students using the POGIL approach are compared to their L-I counterparts under the same instructor, based on the analysis of six semesters of final grades, final exam scores, and final exam questions, the POGIL students are learning more chemistry at a higher learning level.

---

[3]The range in the number of exam questions used is due to variations in content from semester to semester.

Average difference in correct responses

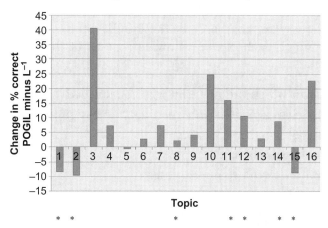

FIGURE 7.4 Of the 16 chemistry topics examined (1–16) on the final exam, overall the POGIL students had more correct responses to the same topics than their L-I counterparts. Some topics did not appear on all the POGIL exams. Asterisks indicate topics that were asked every semester and compared to the L-I group. The topics included a solution problem (1), Lewis structures (2), chiral center identification (3), salt dissociation (4), neutralization (5), acid–base equilibrium (6), radioactive half-life (7), isomerism (8), ionic compounds (9), biological condensation/hydrolysis (10), intermolecular forces (11), functional group identification (12), salt formation (13), biomolecule identification (14), LeChâtelier's principle (15), and physical/chemical property (16).

In addition, because POGIL also emphasizes the learning process, it has been shown to have a positive effect on learning even after students move from a POGIL classroom setting back to a traditional lecture format. This is also true for nursing majors at GSU in POGIL classrooms that have been tracked in their subsequent nursing courses to be better critical thinkers (A. Rushing, pers. comm.).

STUDENT PERCEPTIONS OF LEARNING

Because the POGIL approach incorporates the essential elements of the natural critical learning environment described by Bain, it should enhance student perceptions of learning. Did students perceive higher learning gains under the enhanced classroom environment offered by POGIL? In the allied health course examined, students were asked to complete the Student Assessment of Their Learning Gains (SALG; Carroll et al. 2007) online survey at the end of the semester. This instrument gathers learning-focused feedback from students and asks students to rate on a five-point Likert scale how different components of a course might have helped them to learn. The responses of a cohort of L-I students were compared to the POGIL students. Students were asked how much each of the following aspects of the class helped in their learning:

- **the way in which the material was approached;**
- **how the class activities, lab, reading, and assignments fit together;**
- the pace at which we worked;
- **group work in class;**
- the mental stretch required of us;
- the grading system used;
- the feedback we received;
- **the quality of contact with the teacher;** and
- working with peers outside of class.

Based on the comparison of the two groups, the POGIL students made statistically significant gains ($p = 0.05$) in the boldfaced items. These items reflect an organized, active, critical learning environment that can be created in the classroom using a nontraditional approach like POGIL.

Allied health students can be considered a consumer of chemistry. They are required to apply chemical knowledge to their chosen career, and they need the critical thinking and problem solving skills that go along with full understanding. A judicious selection and integration of topics incorporating biochemistry, a student-centered inquiry-based pedagogy like POGIL, and a classroom environment conducive to learning have been shown here to increase student interest, student learning, and student perception of learning for the nonmajor.

REFERENCES

Bain, K. (2004) *What the Best College Teachers Do*, pp. 99–103. Cambridge, MA: Harvard University Press.

Carroll, S., E. Seymour, and T. Weston (2007) Student Assessment of their Learning Gains (SALG). http://www.salgsite.org (accessed December 2009).

Centra, J. A. and N. B. Gaubatz (2000) Student perceptions of learning and instructional effectiveness in college courses (Research Rep., No. 9, The Student Instructional Report II). Princeton, NJ: Educational Testing Service.

Daly, J. and J. Sarquis (1987) *J. Chem. Educ. 64*, 699–707.

Dever, D. F. (1991) *J. Chem. Educ. 68*, 763–764.

Dolter, K. and L. Marshall (2004) Chemistry in nursing curricula. In Program Book, *Biennial Conference on Chemical Education*, July 18–22, 2004, pp. 223, S433. Ames, IA: American Chemical Society Division of Chemical Education, Iowa State University.

Eberlein, T., J. Kampmeier, V. Minderhout, R. S. Moog, T. Platt, P. Varma-Nelson, and H. White (2008) *BAMBED 36*, 262–273.

Farrell, J. J., R. S. Moog, and J. N. Spencer (1999) *J. Chem. Educ. 76*, 570–574.

Frost, L. D., S. T. Deal, and P. Humphrey (2006) *J. Chem. Educ. 83*, 893–897.

Gagné, R. M. and L. J. Briggs (1974) *Principles of Instructional Design*, 2nd ed. Austin, TX: Holt, Rinehart, and Winston.

Garoutte, M. P. (2007) *General, Organic, and Biological Chemistry: A Guided Inquiry*. New York: John Wiley & Sons.

Hanson, D. (2006) *Instructor's Guide to Process-Oriented Guided-Inquiry Learning*, p. 25. Lisle, IL: Pacific Crest.

Hanson, D. and D. Apple (2004) Process—The Missing Element. http://www.pkal.org/documents/hanson-apple_process–the-missing-element.pdf (accessed December, 2009).

Holme, T. A. (1993) *J. Chem. Educ. 70*, 933–935.

Johnson, D. W., R. T. Johnson, and K. A. Smith (1991a) *Active Learning: Cooperation in the College Classroom*. Edina, MN: Interaction Book Company.

Johnson, D. W., R. T. Johnson, and K. A. Smith (1991b) Cooperative Learning: Increasing Faculty Instructional Productivity. ASHE-ERIC Higher Education Report No. 4, George Washington University, Washington, DC.

Karplus, R. and H. D. Thier (1967) *A New Look at Elementary School Science: Science Curriculum Improvement Study*. Chicago: Rand-McNally.

Keri, G. (2002) *College Student Journal 36*. http://www.findarticles.com/p/articles/mi_m0FCR/is_/ai_95356596 (accessed December 2009).

King, A. (1994) Inquiry as a tool in critical thinking. In *Changing College Classrooms: New Teaching and Learning Strategies for an Increasingly Complex World*, ed. D. F. Halpern, pp. 13–38. San Francisco: Jossey-Bass.

Maher F. A. and M. K. T. Tetreault (1994) *The Feminist Classroom*. New York: Basic Books.

McKeachie, W., P. Pintrich, L. Yi-Guang, and D. Smith (1986) *Teaching and Learning in the College Classroom: A Review of the Research Literature*. Ann Arbor, MI: NCRIPTAL.

Michaelsen, L. K., A. B. Knight, and L. D. Fink (2004) *Team-Based Learning: A Transformative Use of Small Groups in College Teaching*, p. 14. Sterling, VA: Stylus.

Pettigrew, F. and D. Zakrajsek (1984) *Physical Educator 41*, 85–89.

Piaget, J. (1964) *J. Res. Sci. Teach. 2*, 176–186.

Process-Oriented Guided Inquiry Learning (POGIL) Effectiveness of POGIL. http://www.pogil.org/effectiveness/ (accessed December 2009).

Slavin, R. E. (1995) *Cooperative Learning*, 2nd ed. Boston: Allyn & Bacon.

Tracy, H. (1998) *J. Chem. Educ. 75*, 1442–1444.

Timberlake, K. (2004) Why do I need chemistry to be a nurse? In Program Book, *Biennial Conference on Chemical Education, July 18–22, 2004*, pp. S564, 264. Ames, IA: American Chemical Society Division of Chemical Education, Iowa State University.

Treblow, M., J. M. Daly, and J. L. Sarquis (1984) *J. Chem. Educ. 61*, 620–621.

Walhout, J. S. and J. Heinschel (1992) *J. Chem. Educ. 69*, 483–487.

Ward, R. J. and G. M. Bodner (1993) *J. Chem. Educ. 70*, 198–199.

Weimer, M. (2002a) *Learner-Centered Teaching*, pp.99–101. San Francisco, CA: Jossey-Bass.

Weimer, M. (2002b) *Learner-Centered Teaching*, pp. 153–161. San Francisco, CA: Jossey-Bass.

Zoller, U. (1993) *J. Chem. Educ. 70*, 195–197.

CHAPTER 8

Working with Chemistry: A Laboratory Inquiry Program

JULIE ELLEFSON

Harper College, Palatine, IL

INTRODUCTION

Fuming flasks, white lab coats, explosions, toxic substances—these are images that often come to mind when people think of chemists and chemistry. For ordinary people, chemistry is often vastly misunderstood and is often considered too difficult to grasp. Steve Zumdahl (1993) writes that it is "impossible to define (chemistry) concisely because the field is so diverse and because its practitioners perform such an incredible variety of jobs." This statement provides a much broader vision of chemistry than simply "the study of matter." An understanding of chemical concepts is essential not only for chemistry majors but also for other science majors and even students with undecided majors as it leads to a richer understanding of other science disciplines, health fields, and societal issues.

In *The Sciences: An Integrated Approach*, Trefil and Hazen present great unifying ideas or themes in science and include a web of how the great idea has been applied in different science disciplines. In many cases, the application in the different disciplines requires an understanding of basic chemistry. For example, one of their great ideas across the sciences states "Ecosystems, interdependent collections of living

Any opinions, findings, and conclusions or recommendations expressed in this material are those of the author(s) and do not necessarily reflect the views of the National Science Foundation.

things, recycle matter while energy flows through them" (Trefil and Hazen 1998). The concepts listed in the web, acid rain (geology), the ozone layer (astronomy), the food chain (biology), recycling and solid waste (technology), chlorofluorocarbons (CFCs), and the ozone hole (health and safety), carbon dioxide and global climate change (environment), and cleaning smokestacks (physics) all require a chemistry foundation to fully understand. Regardless of the students' areas of study, chemistry helps students develop critical thinking skills—to analyze and evaluate information, to apply concepts to new situations, to think logically, to express ideas orally and in writing, and to provide evidence to support conclusions. Therefore, it is not surprising that many disciplines require at least one semester of chemistry as a prerequisite.

While chemistry professors recognize the beauty and applicability of chemistry, when asked, a student in general chemistry will rarely respond he or she is enrolled in general chemistry "to better understand the nature of matter," or "because an understanding of basic chemical principles and an ability to think critically will make me a better doctor." Instead, most students state that they are enrolled in chemistry because it is required for their major and that the major is typically biology, engineering, pharmacology, nursing, premedical, predental, physician assistant, or other health professions rather than chemistry. At a large urban research institution, only 60–100 students graduate with a degree in chemistry or biochemistry annually. However, 25 programs at that same institution require chemistry, so approximately 2300 students enroll in general chemistry each year. This is not atypical; a majority of the students who take general chemistry will not graduate as chemistry majors.

Students often view their chemistry course as a hurdle to overcome enroute to their desired degree rather than as a foundational course necessary for a richer understanding of their field of interest. In addition, general chemistry students typically have diverse academic backgrounds. Although a year of high school chemistry or a semester of preparatory chemistry serves as a prerequisite for general chemistry at many institutions, the students' high school experiences may be vastly different from one another as distinctions are not made between regular, honors or even advanced placement chemistry.

The same can be said for student high school laboratory experiences. While we may think our incoming college students have already mastered basic laboratory techniques in high school, that is often not the case. In fact, the National Research Council's (NRC) 2006 investigation of the state of laboratory instruction in high schools in the United States concluded "the quality of current laboratory experiences is poor for most students" (NRC 2006). In most high school laboratory programs, students are provided with the question under investigation, the procedure, and data pages. The outcome is known either by the teacher or by both the teacher and students. Students typically work under a time constraint, needing to finish the laboratory in a single laboratory period frequently ranging from 50 minutes to two hours. They are not given time to think about the data, to make revisions to the procedure, or to present their results for discussion and constructive criticism with peers. In short, students arrive on our campuses with little of what we would consider solid laboratory techniques.

Equally disturbing to postsecondary science educators in regard to the student's lack of adequate preparation for college chemistry is that a number of academically capable students are either not pursuing degrees in the sciences (or mathematics and engineering—SME) or dropping out of the fields. The attraction to participate in research-based careers that was strong especially in the 1950s and the 1960s has diminished, and relevance to daily lives has become more important (Lloyd 1992). Traditional lecture-based pedagogical strategies with little collaborative interaction in a competitive environment are also driving students from SME careers (Tobias 1990). Seymour and Hewitt (1994) report the primary reason students switched out of SME majors was because they were turned off to science—they lost interest. Poor teaching, a fast-paced overcrowded curriculum, and inadequate high school preparation are other factors that contributed to the decision to switch. Thus, general chemistry professors are not only challenged to motivate students to recognize and to appreciate the role of chemistry in their desired fields of study and in their lives in general but they are also challenged to provide a stimulating and engaging learning environment to a diverse group of learners. The laboratory is an excellent venue for creating such an environment.

THE ROLE OF THE LABORATORY IN CHEMICAL EDUCATION

Justus von Liebig, professor of chemistry at the University of Giessen, Germany, created the first scientific research and teaching laboratory in 1833 (Michaelis 2003). Other scientists of Liebig's era usually worked alone at home or in libraries so Liebig's laboratory was unique in that it enabled scientists to work together, to converse with others about scientific subjects, and to learn from each other; thus, it provided an excellent environment for students to study and it served as a model for other institutions worldwide.

While this laboratory model is still prevalent in graduate schools, it has not been the model utilized at other educational levels. The role of the laboratory has changed over time, from allowing students to make observations of reactions of elements and compounds and to do simple preparations, to training students to appropriately use apparatuses and instruments while following a standard analytical procedure, to providing students the opportunity to ask and answer testable questions, obtain and analyze data, and draw conclusions based on evidence (Lloyd 1992; NRC 2006). These different roles for the laboratory have resulted in the development of different types of laboratory experiences, all of which are currently utilized in laboratory programs across the United States.

The American Chemical Society (ACS), the NRC, and the American Association for the Advancement of Science (AAAS) have all outlined goals for the laboratory that not only result in skill development but also involve students in the process of science. According to the ACS, in *ACS Guidelines for Chemistry in Two-Year College Programs*, "as an experimental science, chemistry must be taught using appropriate and substantial laboratory work that provides opportunities for open-ended investigations, which promote independent thinking, critical thinking and reasoning, and a perspective of chemistry as a scientific process of discovery" (ACS 2009). In addition, the laboratory experience is "… necessary for students to understand, appreciate, and apply chemical concepts. It should also develop student competence and confidence" (ACS 2009).

In *America's Lab Report*, the NRC (2006) identified the following learning goals attributable to laboratory experiences:

- Enhance mastery of subject matter.
- Develop scientific reasoning.
- Understand the complexity and ambiguity of empirical work.
- Develop practical skills.
- Understand the nature of science.
- Cultivate an interest in science and an interest in learning science.
- Develop teamwork abilities.

Although the AAAS Project 2061 benchmarks are not aimed at postsecondary education, the role of the laboratory described here applies equally well to the college level. As stated in *Benchmarks for Science Literacy*, "Of course, the student laboratory can be designed to help students learn about the nature of scientific inquiry. As a first step, it would help simply to reduce the number of experiments undertaken (making time available to probe questions more deeply) and eliminate many of their mechanical, recipe-following aspects. In making this change, however, it should be kept in mind that well-conceived school laboratory experiences serve other important purposes as well. For example, they provide opportunities for students to become familiar with the phenomena of which the science concepts being studied try to account. Another, more ambitious step is to introduce some student investigations that more closely approximate sound science. Such investigations should become more ambitious and more sophisticated" (Project 2061 1993).

The Working with Chemistry (WWC) laboratory program is an example of an innovative laboratory program that strives to accomplish the goals outlined above and that also provides a context to appeal to interests of the diverse groups of students who enroll in general chemistry.

DEVELOPMENT OF THE WWC PROGRAM

Interdisciplinary collaborative projects and programs are prevalent in research and industry. *Chemical & Engineering News* reported that many companies expect their professional employees to participate on cross-functional teams in which they collaborate with colleagues from other disciplines (Ainsworth 1999) and such projects are becoming more widespread in academia as well. For example, nanotechnology and neuroscience programs enlist faculty from a variety of disciplines, including chemistry, biology, engineering, and medicine (AND&P 2009; NIN 2009). The National Science Foundation (NSF) has a separate category for interdisciplinary projects in the Course, Curriculum and Laboratory Improvement program, and additional funding is available if two- and four-year institutions collaborate on projects. Additionally, the NSF Advanced Technological Education program requires collaboration by multiple science, technology, engineering, and mathematics (STEM) fields for a successful proposal (NSF, Undergraduate Education [DUE] 2009).

The WWC laboratory program, developed under the auspices of an NSF grant (DUE No. 96-53080), was designed to provide explicit examples of how chemistry is used by professionals who are not educated and trained as chemists per se. The program serves as an excellent example of an academic interdisciplinary project that was successful as it was the result of the efforts of a network of faculty in different disciplines and from different institutions. The primary goal of the WWC laboratory program is to develop an understanding of how chemistry gathers evidence and solves problems. In accomplishing this goal, students should also develop a mastery of techniques and ideas so that they can gather and interpret data independently. In addition, they should be able to articulate, with reference to the written materials and personal experiences, how someone in a real work environment might find a particular technique or idea useful in solving a problem.

The professional faculty involved in developing the scenarios that serve as the basis for the experiment groups were all faculty at a four-year research institution, while the network of faculty from three community colleges performed, edited, and class tested the experiments. The benefits of the network extended beyond developing improved laboratory materials appropriate in different instructional settings. The members of the network submitted and were awarded an NSF Instrumental and Laboratory Improvement grant for UV–vis spectrophotometers. The network also fostered the exchange of ideas between faculty in different disciplines and at different institutions, and it established a relationship between institutions resulting in collaboration on other projects.

FORMAT

The WWC program consists of experiment groups that each initially included a scenario and a set of three experiments—a skill-building laboratory, a foundation laboratory, and an application laboratory. The scenario presents a situation or problem that nonchemistry professionals, such as an ecologist or a nurse, may encounter in their field. Examples of four experiment groups are provided later in this chapter.

It is in the application laboratory that the students address the problem presented in the scenario. The procedure is generally not written in detail; no data pages are included, and the students do not know the outcomes of the experiment in advance. However, as explained earlier, first-year chemistry students typically do not possess well-developed laboratory skills nor do they usually have the knowledge base to fully plan their own procedures. The WWC experiment groups initially included a skill-building laboratory and a foundation laboratory to enable students to learn basic techniques and strategies, such as basic titration skills or the generation of a calibration curve, that they could then utilize in the application laboratory. The program was revised and each experiment group was reduced to two experiments—skill building and application—because general chemistry is such a content-dense course that it is difficult for programs to commit 3 weeks of a 16-week semester to a single topic and a set of laboratory skills.

In all of the experiments, students are provided the opportunity to develop individual skills and to communicate their own results. All students are expected to learn and to practice all laboratory techniques utilized in each experiment. Students should also each record all data in their own notebooks. Since no data pages are provided, they need to decide what to record and how to organize their notebooks. The experiments are designed so students submit individual final reports. While each experiment includes a "report" section, outlining for students what questions to consider as they analyze their data and, in the application laboratory, address the question or issue posed in the scenario, the format of the report is not specified. The application laboratories lend themselves well to formal reports in which students write an abstract and include introduction, experimental, results and discussion, and conclusion sections. Rather than submitting their reports to the instructor, they could be shared with peers not in their group who would review the papers, add comments, and return them to the original author, thus giving the students experience with a peer review process. The student could then revise the report based on peer feedback and submit it for final approval (and a grade) to the instructor. Alternatively, the students could prepare posters and the instructor could schedule a poster session to which other faculty or students could be invited. Group reports are another option. As explained below, each experiment has a strong collaborative group component, so if students work in groups of three and stay with the same group for three different experiment groups, each student could assume the role of lead writer for the laboratory report. The lead author would submit a draft to the group members who would edit and return the report to the lead author who would then prepare the final document. The drafts would be submitted with the final document so the instructor could ensure all students participate.

The development of collaborative skills is another major objective in the WWC laboratory program; therefore, each experiment group contains a team component. Instructors can either assign students to groups or allow the students to self-select their team members. If students self-select, they will likely choose to work with friends, which may or may not result in a productive collaboration. Assignment of groups forces students to learn to work with someone they may not know, an experience they are likely to encounter outside of the classroom. Early in the semester, the instructor may not know the students well enough to be able to divide them into groups with mixed skill levels, so random assignment is a valid strategy. Changing the groups from one experiment group to another for the first few weeks can assist the instructor in determining the most effective groups. Direct observation and group evaluations completed by the students can be utilized to assess the effectiveness of the groups. Alternatively, after working with different people for the first few experiment groups, students can then be allowed to self-select their groups for the remainder of the semester. If permanent groups are assigned at the beginning of the semester, students can be given the opportunity to file complaints with the instructor if they have evidence the group is not functioning, resulting in either intervention strategies to assist reconciliation of the group or the removal of a group member—either the one who filed the complaint or the offending member. In this case, another group would have to allow a new member to join. Generally, teams should consist of no more than four members. It is too easy for someone to disengage and simply to watch the rest of the team if the team is too large. Once the teams are established, the members need to decide roles. It is important for the laboratory instructor to observe how the members are interacting, to intervene to prevent the dominant personality from becoming a dictator, and to ensure active participation by all members. Team members collaborate in order to develop or to refine a procedure, to share data, to prepare valid samples for each other, and to discuss results.

Since each experiment has chemistry concepts at its core, students will develop a deeper understanding of the subject matter addressed in the experiment group such as buffers, stoichiometry, solution chemistry, kinetics, thermochemistry, and equilibrium, especially if they are allowed to struggle with the application laboratory rather than have the instructor give them the answers, and if they are given time to discuss their results and conclusions with their peers. In their review of the laboratory in science education, Hofstein and Lunetta (2004) cited works that indicate that if students perceive investigations to be meaningful and connected to their own experiences and are allowed to reflect on their work and engage in meaningful dialogue with their peers, they can begin to develop more scientific concepts.

It is imperative students learn to work safely in the laboratory so every experiment group highlights any necessary safety precautions. Chemicals are selected to minimize risk and students are expected to discuss their experimental design with the instructor to be sure it is appropriate and safe.

INQUIRY BASED

What does it mean that WWC is a laboratory inquiry program? What does inquiry mean? Science education reform efforts have placed an increasing emphasis on making inquiry a central component of science programs. The NRC National Science Education Standards state, "Students will engage in selected aspects of inquiry as they learn the scientific way of knowing the natural world, but they also should develop the capacity to conduct complete inquiries," and they define inquiry as "a multifaceted activity that involves making observations; posing questions; examining books and other sources of information to see what is already known; planning investigations; reviewing what is already known in light of experimental evidence; using tools to gather, analyze, and interpret data; proposing answers, explanations, and predictions; and communicating the results. Inquiry requires identification of assumptions, use of critical and logical thinking, and consideration of alternative explanations" (NRC 1996).

As stated in *Benchmarks for Science Literacy* of Project 2061 (1993), "scientific inquiry is more complex than popular conceptions would have it. It is, for instance, a more subtle and demanding process than the naive idea of 'making a great many careful observations and then organizing them.' It is far more flexible than the rigid sequence of steps commonly depicted in textbooks as 'the scientific method.' It is much more than just 'doing experiments,' and it is not confined to laboratories. More imagination and inventiveness are involved in scientific inquiry than many people realize, yet sooner or later strict logic and empirical evidence must have their day. Individual investigators working alone sometimes make great discoveries, but the steady advancement of science depends on the enterprise as a whole." Colburn (2000) describes inquiry as "the creation of a classroom where students are engaged in essentially open-ended, student-centered, hands-on activities."

Educators are challenged to first interpret the myriad of definitions/explanations of inquiry and then to design an inquiry-based learning environment. Regardless of how inquiry is defined, the laboratory is an excellent setting for inquiry-based learning. According to Domin (1999), inquiry is one of four classifications for laboratory instruction styles; the other three are expository, discovery, and problem-based (Table 8.1). These styles differ in whether the outcomes are predetermined or undetermined; the approach is inductive (specific to general) or deductive (general to specific), and the procedure is provided or student-generated.

Colburn (2000) differentiates between different levels of inquiry labeled as structured, guided, and open, dependent upon which of the following—problem, materials, and procedure—are provided for the students (Table 8.2). The structured inquiry is similar to Domin's expository style except that the laboratory experiences are less governed in

TABLE 8.1 Laboratory Styles (Domin 1999)

Laboratory Style	Outcome	Approach	Procedure
Expository	Predetermined	Deductive	Given
Inquiry	Undetermined	Inductive	Student generated
Discovery	Predetermined	Inductive	Given
Problem based	Predetermined	Deductive	Student generated

TABLE 8.2 Type of Inquiry (Colburn 2000)

Type of Inquiry	Problem	Materials	Procedure
Structured	Provided	Provided	Provided
Guided	Provided	Provided	Student generated
Open	Student generated	Student generated	Student generated

TABLE 8.3 Level of Openness in Laboratory Exercises (Schwab 1962; Herron 1971)

Level of Inquiry	Problem	Ways and Means	Answers
Level 0	Given	Given	Given
Level 1	Given	Given	Open
Level 2	Given	Open	Open
Level 3	Open	Open	Open

terms of what students are to observe and to record as data, and the outcomes are not predetermined.

Herron (1971) and Schwab (1962) characterize inquiry in experiments according to the level of openness of three characteristics—problem, ways and means, and answers (Table 8.3).

Finally, Buck et al. (2008) built upon Herron's work and developed an expanded rubric to characterize the level of inquiry in laboratory activities or exercises (Table 8.4).

WWC experiment groups were primarily rated as structured inquiry when evaluated using the scheme developed by Buck et al. In Herron's system, most would be rated level 1 or level 2 with respect to the level of openness. While the problem and, in many cases, part or all of the procedure is provided to students, they are responsible for collecting and interpreting data in order to answer the posed testable question. In addition, some of the experiments do not have a predetermined outcome. For example, in the *Ecological Element Cycles* experiment group, students may bring in their own soil to analyze so the nitrogen content is truly an unknown. Different types of leaves and the conditions under which the compost jars are stored in the *Acid-Base Titration and the*

TABLE 8.4 Levels of Inquiry (Buck et al. 2008)

Characteristic	Level 0: Confirmation	Level ½: Structured Inquiry	Level 1: Guided Inquiry	Level 2: Open Inquiry	Level 3: Authentic Inquiry
Problem/question	Provided	Provided	Provided	Provided	Not provided
Theory/ background	Provided	Provided	Provided	Provided	Not provided
Procedures/ design	Provided	Provided	Provided	Not provided	Not provided
Results analysis	Provided	Provided	Not provided	Not provided	Not provided
Results communication	Provided	Not provided	Not provided	Not provided	Not provided
Conclusions	Provided	Not provided	Not provided	Not provided	Not provided

Global Carbon Cycle group will result in different rates of decay, and therefore different conclusions may be reached. Students are also expected to determine if their data are acceptable and, if not, to make adjustments to their procedure and to collect more data. For example, if, when they are generating a calibration curve to determine the iron ion concentration in a simulated blood sample, their data result in a low correlation coefficient, they need to determine if this is due to poor experimental technique or if the nonlinear trend is real. Finally, they are expected to make connections between the scenario, the data they collected, and the chemical principle underlying the experiments. These connections are best forged if the students engage in a discussion of their results with the instructor facilitating the discussion; otherwise, it is not likely the students will tie the experiment group together. The post-laboratory discussion times are excellent opportunities for students to present their results orally and for their peers to review their work. Students can then incorporate the feedback received into their final reports. This sharing of results, peer review, and subsequent revision process serves as a model for the students as to how scientists work.

IMPLEMENTATION: SUCCESSES AND BARRIERS TO SUCCESS

The person responsible for students in the laboratory setting can have a profound impact on what the students experience during the laboratory period. Experiments that are more student-centered and inquiry-based (regardless of the way inquiry is defined), that do not have a step-by-step procedure and fill-in-the-blank data pages, are much more labor intensive for the laboratory instructor. They need to be able to guide students, to promote critical thinking, and to avoid just giving the students answers. The laboratory instructor may be the professor, either full-time or adjunct, or a teaching assistant. Teaching assistants often view themselves as helpers, not as teachers, and since they typically have their own classes to attend and research to do, they may try to get their students through the material and out the door as quickly as possible. Many students seem happy with this laboratory arrangement, at least initially. In the implementation stage of the WWC laboratory program at a four-year research institution, the teaching assistants were not given any special preparation in advance and received training only in their weekly meetings with individual instructors. The students were expected to merely "catch on" to a new way of doing laboratories. Thus, many teaching assistants continued in their traditional mode of supervision and told the students everything they needed to do, eliminating the student's involvement in the design of the procedure. Other teaching assistants thought their new role was to never answer a question, to leave the students completely on their own, and just to make sure they did not do anything dangerous. This clearly was not an appropriate or effective method of training teaching assistants.

The second year the WWC laboratory program was used at the same four-year institution, the teaching assistants underwent a more rigorous training program prior to the start of the semester. The teaching assistants actively participated as if they were students and performed some of the laboratories they were expected to teach that semester. The professor running the training session modeled appropriate behavior for an instructor teaching in this laboratory program, and the personal experiences the teaching assistants gained enabled them to better understand what their students would be going through and to be able anticipate questions they were likely to receive. This type of rigorous, hands-on training produced much more prepared and effective laboratory instructors. Roehrig et al.'s (2003) work supports the conclusions reached following the WWC initial implementation phase, namely, teaching assistants need specific training to teach inquiry-based experiences more effectively.

To make WWC or any inquiry-based laboratory experience effective, the student-centered, nonexpository methodology needs to be introduced early in the laboratory program and preferably in the lecture portion of the class as well. One cannot drop one inquiry-based experiment or activity into the curriculum and expect students to feel confident and to progress satisfactorily during that experience. Students gain confidence throughout the semester, the more they are expected to be involved in design, interpretation, and discussion of results. Instructors must avoid giving students the answers but must learn to ask students questions to help them progress. The instructor must remain very actively engaged with the students in the laboratory. Finally, students must be given time to discuss the results with each other and in the presence of the instructor.

When students first start working in a more open-ended, student-centered laboratory environment, they typically have many more questions than in a traditional laboratory, and they are almost afraid to make a move without first consulting the instructor. They want to know the "right" way to proceed. As students continue to work in such a setting throughout the semester, many exhibit increased confidence and independence.

Another difficulty associated with successful implementation is when a large number of faculties are involved in teaching the laboratories. A community college utilizing some of the WWC experiment groups phased out the application laboratories from its general chemistry curriculum primarily because the instructors believed they were very difficult for the students and could not be attempted without the skill-building laboratories. However, they did not want to spend 2 or 3 weeks on the same concepts. Chemistry classes at this community college are limited to 24 students and the same professor is responsible for all aspects of the class with that group of students. In a typical semester, the chemistry department offers 10 or 11 sections of general chemistry with a minimum of seven different instructors. The faculty assigned to teach general chemistry varies from semester to semester, and it is not uncommon to have one or more new professors teaching the course in any given semester. In order for the program to be effective, the professors must be committed, as they play a very different role in the student-centered laboratories than in traditional expository laboratories. It is not easy to initiate change unless the faculty involved perceive a problem with the status quo and are committed to make a change. If, as the chemistry department at this or any other institution assesses their general chemistry curriculum and laboratory goals, they make a decision to adopt experiments like the WWC experiment groups, they would potentially have greater active participation by all faculty if they met to discuss the reasons behind the change and scheduled training sessions similar to those used by the research university to assist their teaching assistants to teach effectively in a student-centered laboratory environment.

EVALUATION

The evaluation of the WWC laboratory program initially consisted of direct observations of students and their professors working in the laboratory and interviews with groups of students. The students were asked if they were making the connection between their fields and chemistry, if the group work was effective, how the WWC laboratories compare with traditional laboratories, and what could be done to improve the program. The study indicated most students did not make the connection between chemistry and other fields from which the experiments are drawn unless that connection was made explicit by a discussion of the scenario before they started the experiment group and after they finished collecting data in the application laboratory. However, students did recognize the applicability of working collaboratively with others. The following quote from a general chemistry student summarizes these findings:

> I don't know how it (the chemistry) will apply to me in the future, but absolutely the real life … you don't walk into work and there's your procedure for you. You meet with your co-workers … you discuss your problems and come up with a plan of action. Since most of the labs were group oriented … and the procedure wasn't spelled out for you … to make us work together and think about it.

Students generally found the groups to be effective in that they could share ideas, help each other understand, and divide the labor. However, they also recognized how a lack of preparation by one or more members was detrimental to the group's productivity.

The primary difference the students noted between the WWC laboratories and the laboratories to which they were accustomed was the lack of step-by-step directions, and many students struggled with this and the absence of data pages. As one student succinctly stated, "They're trying to get you to think more independently." Another stated, "There was a lot more background information as to why we were doing this … I felt like we really had to figure out what we were doing on our own."

As for improvements, the students wanted more discussion time before and after completing the experiment. For most of the students, this type of laboratory program was a completely new experience, and they were being stretched out of their comfort zone; they needed to be more engaged with their instructor and peers in order to develop confidence and to regain some level of comfort with the work. Additionally, as noted earlier, the time to reflect and to interact with peers is important for students to enhance their conceptual understanding.

While the original assessment plan for this program did not include an assessment of conceptual understanding, one of the adopters of the program has determined that students who have participated in the application laboratory perform statistically better on quizzes than students who have not participated (survey results). Since each group of students is different, it is important for each institution to develop its own assessment plan and to evaluate its own laboratory program. Direct observation of students in the laboratory reveals how much each student contributes to the completion of the experiments and whether they are developing laboratory skills and safe practices. Evaluation of the written laboratory reports and oral presentations reveals how well students can manipulate the data, draw appropriate conclusions, and understand the ambiguity of empirical work. Laboratory practical exams in which students are expected to design a procedure to answer a posed question can determine whether students learned to think independently and to problem solve. The last 2 or 3 weeks of the semester can be reserved for a completely open-ended project to determine the extent to which students can fully plan, carry out, and evaluate an investigation. Institutions should design appropriate assessment strategies to match their laboratory goals.

EXAMPLES AND ADAPTATIONS

Many students enrolled in general chemistry plan to pursue a pharmacy degree or a degree in a health profession, so scenarios drawn from these fields appeal to a large percentage of students. In *Metal Ions and the Blood: Treat Iron Deficiency ... and Overload*, an experiment group developed with a professor from the department of medicinal chemistry in a college of pharmacy, students assume three roles as presented in the scenario: a technician working in a clinical laboratory, a doctor, and a pharmacist. "Two blood samples have just arrived in the laboratory. One is from a woman who has been experiencing fatigue, weakness, shortness of breath, low blood pressure, and headaches. The doctor suspects iron-deficiency anemia. Another is from a child who has taken several of his mother's iron supplement pills, thinking they were candy." You receive one of the two samples. Your job is to measure the serum iron level in your sample and to report the results to the doctor and the pharmacist. They would then prescribe and dispense the necessary treatment to raise the iron level of the woman and to lower that of the child. You will then temporarily assume the role of the doctor and decide how to "treat" the patient either by increasing or decreasing the iron concentration in your sample. Next, you assume the pharmacist's role by dispensing the treatment before returning to a technician role to assess the final results (Wink et al. 2005).

The skill-building laboratory introduces students to spectrophotometry, the technique used to analyze iron in blood. They also prepare serial dilutions and explore the relationship between absorbance and concentration to determine the concentration of an unknown solution. Then, in the application laboratory the students are told they need to generate a calibration curve, but they are not given detailed directions as to how to prepare the solutions. Once they have determined their calibration curve is acceptable (data points are consistent and reproducible), they obtain an unknown sample—the simulated blood sample from a patient whose iron level is either too high or too low. After measuring the concentration in the sample, they need to decide how much iron to add or to remove so the patient's blood iron level is back in the normal range. After "treating" their sample, they again measure the iron concentration to determine if the treatment was successful. Students work collaboratively to generate the calibration curve. To better ensure students develop individual laboratory skills, each could receive his own "patient" sample to analyze. The group can decide upon a treatment plan, but each student would be responsible for carrying out the treatment and for determining the effectiveness of the treatment by remeasuring the iron concentration. To save time, standards of iron solutions with concentrations in the normal range for females and children are available for use in generating the calibration curve. Since the samples are outside of the normal range, students are required to prepare two additional solutions, one with a greater concentration and one with a lower concentration than those in the normal range, thus extending their calibration curve and avoiding the need to extrapolate to determine the concentration of the unknown. If time is not a factor, students can be responsible for preparing all of their standards from a stock solution. They would then have to research what the normal serum iron concentrations are for children and for adult females in order to determine how to treat their "patient" samples and whether the treatment was effective.

In *Acid-Base Titration and the Global Carbon Cycle*, the scenario and the application laboratory were developed in conjunction with a field ecologist who simply stated, "We can have the students set up in the laboratory what we have done in the field. They just have to set up an apparatus that traps the leaves and collects the CO_2." This experiment provides an example of a real connection between chemistry, biology, and ecology for the students and addresses a societal issue prominent in their daily lives. The scenario sets the students up as summer interns working for a biologist who wants them to use their knowledge of chemistry to help predict the effects of rising atmospheric CO_2. The biologist "tells you that the CO_2 concentration in the atmosphere is rising because of the burning of petroleum, coal, and wood. She says that predicting the effects of an atmosphere with a high CO_2 concentration on society is important. It is possible to imagine, she says, negative effects such as climate warming, leading to sea-level rise and coastal flooding. But, she continues, it is also possible to imagine positive effects such as improved plant growth and more land in the north available for agriculture. The biologist shows you the equation

$$C_n(H_2O)_m + n\,O_2 \rightarrow n\,CO_2 + m\,H_2O + energy$$

She grows plants in chambers containing elevated partial pressures of CO_2 in order to study the right to left reaction. She suggests that you study the left to right reaction, which gives the amount of CO_2 returned to the atmosphere by decomposing leaves. She says there is great interest in whether some of the carbon fixed by plants fails to decompose and is thus kept out of the atmosphere. And with that, she goes back to her own work.

But what exactly should you do? You decide to spend some time in the library reading about the effect fossil fuel use has on the carbon cycle. You learn two things. First 90% of the carbon returned to the atmosphere is from the respiration of decomposers (bacteria and fungi) breaking down dead leaves. Second, the burning of fossil fuels releases other substances in addition to CO_2. One of these is nitrate, which, like CO_2, is a plant nutrient. You decide to do two experiments that will answer the biologist's question and a question formed in your mind as you read in the library. First, the biologist asked whether some of the carbon fixed by plants fails to decompose. You do not have time to study wood decomposition, and you cannot visit peat bogs where carbon is probably accumulating. So you plan to use leaves and ask whether they could decompose in a year under favorable conditions. You reason that if the number of days it would take leaves to decompose under favorable laboratory conditions is more than a year, then carbon would accumulate and there would be less in the atmosphere.

Your library reading made you curious about a second question that your biologist supervisor did not say anything about. The nitrate that is released from burning fossil fuels is an important plant nutrient.

Could nitrate increase the decomposition rate? If so, that would be a reason to expect rapid decomposition of leaves and little accumulation of dead plant material (Wink et al. 2005).

In the application laboratory, students determine the quantity of carbon dioxide produced from the decay of leaves over a given period of time and report their results as milligrams of CO_2 per gram of leaf per day. They set up compost jars and a control jar a week in advance to give the leaves a chance to decay. The students back titrate to determine the number of moles of sodium hydroxide left unreacted, which enables them to ultimately determine the quantity of carbon dioxide produced from the decay of the leaves. The previously done skill-building laboratory has given them experience with back titrations when they analyzed the number of moles of hydrochloric acid neutralized by different commercial antacids.

While the questions are posed for the students in the scenario, and the procedure for setting up the compost jars is provided for the students, they need to determine what reagent will selectively precipitate the carbonate ions produced when the carbon dioxide reacts with the sodium hydroxide, and they have to decide how they will treat their compost jars after performing practice titrations on a mixture of sodium hydroxide and sodium carbonate with and without precipitating out the carbonate ions. They also need to evaluate their results, determine whether or not the leaves will likely decay in a year, and analyze atmospheric carbon dioxide concentration data collected at Mauna Loa, Hawaii. The experiment can be modified to be more of an open inquiry laboratory as defined by Buck et al. (2008). Students do not have to be told how to set up the compost jars. They can be provided with the complete scenario and the information that sodium hydroxide absorbs carbon dioxide producing sodium carbonate and water. It would then be their responsibility to figure out how to "trap" the carbon dioxide. Students can also ask questions as to what impact the conditions under which the compost jars are left will have on the rate of decay of the leaves. For example, does the amount of water with which the leaves are wet, the temperature, light conditions, or type of leaf have an impact on the rate of decay? They can then set up appropriate compost jars that will enable them to answer the question they pose. The myriad results generated in this laboratory certainly introduce students to the complexity and ambiguity of empirical work.

Thermochemistry: Materials and Temperature Control introduces the students to mechanical engineering by setting them up in the scenario as employees in a company that specializes in the design and manufacture of materials used in fireproofing. This experiment is unique in that both the skill-building laboratory and the application laboratory address aspects of the scenario.

The students are given two assignments. First, they are asked "to develop a workable blueprint for the manufacture of small, fire-resistant boxes. What to do you need to consider? To answer this, you must first bear in mind the purpose of making a safe fireproof. Obviously, you do not want the materials inside to melt or to burn. This means you must control the flow of heat between the fire on the outside and the contents of the safe. Most fireproof safes are constructed by putting a smaller box inside a larger box. The filler (or interstitial) material between the two boxes must prevent the heat of the outside fire from damaging the contents inside the safe. The ultimate question, then, is to identify a reasonable interstitial material." Another important fireproofing application is in protecting high-rise buildings. The World Trade Center disaster provides a shocking example of the devastation that can result from a runaway chemical fire. "Although the WTC buildings were initially weakened by the impacts of the planes, it was the resulting fires that caused the structures to collapse. For high rise buildings such as these, city fire codes require fireproofing that will contain a fire for approximately two hours. This generally allows time to evacuate occupants, salvage possessions, or quench the fire. In the case of the WTC, however, evidence ultimately showed the impacts had knocked some of the fireproofing material off the steel supports. Your second task, then, is to find a material that will absorb enough heat to fireproof structural materials" (Wink et al. 2005).

Students are introduced to calorimetry in the skill-building laboratory and immediately apply what they learn to address one of the two tasks outlined in the scenario, namely, to determine an appropriate interstitial material for a fireproof safe. They first need to construct a calorimeter. Several different calorimeters are described, but materials can be placed out in the laboratory with no further directions, and students can decide what to use. Next, they are instructed to determine the heat capacity of their calorimeter and then the specific heat capacity of copper, cork, glass, and concrete. The copper analysis serves as a check for the precision and accuracy of the measurements obtained using their calorimeter. If different groups constructed calorimeters using different materials, posting class results will enable the students to determine if one calorimeter design is more effective than another. Students are provided with the equation to calculate the temperature inside a safe during a fire. One of the parameters in this equation is the mass of the interstitial material. Design specifications for a fireproof safe are also included in the background section of the experiment group so students can calculate the mass of the material needed to fill the interstitial volume (which they calculate) if they know the density of the materials being considered. Thus, they are instructed to determine the density of each material for which they determine a specific heat capacity. They are not, however, given explicit instructions for measuring density.

While the procedure for this experiment is provided with a fair amount of detail, students can have input into the design of the calorimeter and, although a list of materials is provided, different materials can be made available and they can choose the ones they think would be most effective. The inquiry aspect of this experiment lies primarily in the detailed analysis of their results required to suggest an appropriate interstitial material and the need to assess sources of error. If, for example, students have a large error in the determination of the specific heat capacity of copper, they must decide what could contribute to the error and then try to redesign their calorimeter or their technique in order to limit the error.

Students design a calorimeter again in the application laboratory, using what they learned in the skill-building laboratory to design as effective a calorimeter as possible using the available materials. Then they practice determining the heat of reaction of the decomposition of hydrogen peroxide. Finally, they work with their group to design and test a plan to create a heat sink that will absorb and retain some of the evolved heat generated from the reaction. Again, students are expected to analyze the effectiveness of their plan and to determine major sources of error. Since they are instructed to make some assumptions, namely, that the specific heat capacity of the hydrogen peroxide solution is the same as that of pure water and that the mass of the water in solution is the same as that of the entire peroxide solution, they can include an assessment of the impact of these assumptions on their results or they can determine a way to eliminate these assumptions from their final calculations.

Health career programs require students study chemistry so the *Buffers and Life: Save a Cardiac Arrest Patient* especially appeals to those interested in medical careers. Students are introduced to an emergency room situation in the scenario, outlining a need to understand buffers. "In a hospital emergency room, a patient has just been wheeled in following a cardiac arrest in his home. The paramedics were able to restart his heart after a couple of minutes, and prior to that, CPR was administered so that enough oxygen flowed to his tissues to keep him alive. But his face has the ashen color of someone close to death, and his skin is cold. Though his heart is beating again, he is in danger of death from shock. Immediately, one emergency room nurse starts to prepare an intravenous tube, and another gets a bag of fluid from the supply cabinet. The doctor says, 'Yes, let's start him on bicarb' and the nurses begin to administer the fluid, one that helps to restore the victim's blood to its normal state and staves off further damage. The pH of the blood of healthy adults is usually between 7.35 and 7.45. When the body starts to become acidotic, blood pH drops as the body is unable to remove carbon dioxide when the blood stops circulating in cardiac arrest. Blood pH can also rise (become alkalotic) if vomiting occurs and the body uses up carbonic acid to reestablish the stomach acid. In these cases, there is often enough time for a corrective measure. Most cases of blood pH imbalance are treated by allowing, or inducing, the body to make the necessary adjustments. This process is known as compensation by the blood. However, certain events cause such dramatic and threatening pH changes that we cannot wait for natural compensation. When a cardiac arrest occurs, the blood stops moving through the body. Immediately, the concentration of carbonic acid starts to rise, lowering the pH. In addition, normal metabolism halts and lactic acid, the product of partial metabolism of carbohydrates, starts to accumulate. This, too, causes the pH to drop. Even after the heart is restarted, the change in the blood's buffer may be so severe that it places the patient in grave danger. Administration of hydrogen carbonate is sometimes the only way to prevent further, perhaps fatal, damage from occurring. Consequently, an understanding of blood pH and buffers is essential to proper methods in the emergency room. In these experiments, you will progress from studying how buffers work to the point where you are ready to simulate the time-sensitive nature of an emergency room procedure. The material you learn in week one will be essential if you are to save your 'patient' in week two" (Wink et al. 2005).

Before students design and prepare their own buffer solution in the skill-building laboratory, they calibrate a pH meter and study the characteristics of a prepared buffer solution. Their ultimate goal is to determine the capacity of their buffer. Although the scenario describes the carbonic acid–hydrogen carbonate buffer system in the body, this system is difficult to study in the general chemistry laboratory due to the ease with which carbon dioxide is lost to the room atmosphere. Thus, in the application laboratory, students work with a similar buffer system—dihydrogen phosphate/hydrogen phosphate. Before they are challenged to "save a patient" in the application laboratory, the students demonstrate that they can disturb and restore a buffer solution to an expected pH range. Next, they use the techniques learned in the skill-building experiment to characterize a "normal" blood buffer. Third, working with their group, they develop a treatment protocol on a simulated patient by determining the amount of hydrogen phosphate needed to restore a sample of the "patient's blood" that has a pH that is too low for a healthy patient. Next, each student obtains a "patient" from the instructor, measures the pH of the sample, and calculates how much hydrogen phosphate they need to add to restore the pH to 7.35–7.45. They each have a maximum of 30 minutes to figure out how to restore the pH and "save the patient." Finally, they prepare the necessary volume of hydrogen phosphate and add it all at once to the "patient" sample. The instructor measures the pH of the "treated patient" and notes whether or not the pH is in the target range. This experiment provides students with the suspense of working in an emergency room without the risk associated with working with real patients.

CONCLUSION

The WWC laboratory program was designed in part to give general chemistry students an opportunity to experience chemistry as it relates to other disciplines and to develop a better understanding and appreciation of the role of chemistry in their careers and lives. The experiment groups can also serve as a springboard into deeper investigations of the problems presented in the scenario; for example, the leaf laboratory can be conducted in conjunction with a study of greenhouse gases and global climate change. Finally, the program can also be used as a model for programs to develop their own scenarios and application laboratories. This could lead to the establishment of an interdisciplinary network, potentially fostering relationships that extend beyond the initial project that further enhances the learning environment for the students.

REFERENCES

American Chemical Society (ACS) (2009) *ACS Guidelines for Chemistry in Two-Year College Programs*, pp. *11*, 13–14. Washington, DC: American Chemical Society.

Ainsworth, S. J. (1999) *C&EN 77*(46), 54–59.

Association of Neuroscience Departments and Programs (AND&P) (2009) http://www.andp.org/ (accessed June 2009).

Buck, L. B., S. Lowery Bretz, and M. H. Towns (2008) *J. Coll. Sci. Teach. 38*, 52–58.

Colburn, A. (2000) Science Scope, March, 42.

Domin, D. S. (1999) *J. Chem. Ed. 76*, 543–547.

Herron, M. D. (1971) *School Review 79*, 171–212.

Hofstein, A. and V. N. Lunetta (2004) *Sci. Ed. 88*, 28–54.

Lloyd, B. W. (1992) *J. Chem. Ed. 69*, 633–636.

Michaelis, A. R. (2003) *Interdiscip. Sci. Rev. 28*, 280–286.

National Institute for Nanotechnology (NIN), Northwestern University, Evanston, IL (2009) http://www.iinano.org/content/Faculty/index.htm (accessed June 2009).

National Research Council (NRC) (2006) In *America's Lab Report*, eds. S. R. Singer, M. L. Hilton, and H. A. Schweingruber, pp. 6, 53. Washington, DC: National Academy Press.

National Research Council (NRC) (1996) *National Science Education Standards*, p. 23. Washington, DC: National Academy Press.

National Science Foundation (NSF), Undergraduate Education (DUE) (2009) http://www.nsf.gov/div/index.jsp?org=DUE (accessed June 2009).

Project 2061 (1993) *Benchmarks for Science Literacy*, p. 9. New York: Oxford University Press.

Roehrig, G. H., J. A. Luft, J. P. Kurdziel, and J. A. Turner (2003) *J. Chem. Ed. 80*, 1206–1210.

Schwab, J. J. (1962) The teaching of science as inquiry. In *The Teaching of Science*, eds. J. J. Schwab and P. F. Brandwein. Cambridge, MA: Harvard University Press.

Seymour, E. and N. M. Hewitt (1994) *Talking About Leaving*, p. 54. Boulder, CO: Bureau of Sociological Research.

Tobias, S. (1990) *They're Not Dumb, They're Different*. Tucson, AZ: Research Corporation.

Trefil, J. and R. M. Hazen (1998) *The Sciences: An Integrated Approach*, p. 473. New York: Wiley.

Wink, D. J., S. Fetzer Gislason, and J. Ellefson Kuehn (2005) *Working with Chemistry: A Laboratory Inquiry Program*. New York: Freeman.

Zumdahl, S. S. (1993) *Chemistry*, p. 1. Lexington, MA: Heath.

CHAPTER 9

Making Chemistry Relevant to Science and Engineering Majors

JULIE K. BARTLEY,[1] SHARMISTHA BASU-DUTT, VICTORIA J. GEISLER, FAROOQ A. KHAN, and S. SWAMY-MRUTHINTI

University of West Georgia, Carrollton, GA

[1]Present address: Geology Department, Gustavus Adolphus College, St. Peter, MN

Most students enrolled in general chemistry courses during their freshman year at the University of West Georgia (UWG) want to major in either "premed," "pre-pharmacy," or "pre-engineering" and do not express an interest in pursuing a career in chemistry. These students struggle simultaneously with the challenges common to all beginning college students (independence, poor time management, level of rigor, financial issues, etc.) as well as the significant challenge of assimilating content from multiple, simultaneous science (one year of general chemistry followed by one year of physics and/or biology) and mathematics courses. As a means to increase the number of students seeking and receiving baccalaureate degrees in science, technology, engineering, and mathematics (STEM) at UWG, several experienced faculty members tried to identify barriers that may exist for our STEM students and to develop strategies that address these barriers.

First, we found that offerings in introductory mathematics and science courses often make little effort to connect with one another. Second, these introductory courses seldom make connections to the real world, particularly within students' identified fields of interest. Third, incoming students have limited familiarity with the career

Any opinions, findings, and conclusions or recommendations expressed in this material are those of the author(s) and do not necessarily reflect the views of the National Science Foundation.

opportunities for science majors, which they narrowly perceive as comprising doctors, dentists, pharmacists, and engineers. Even for those students who have some idea of a career path, they are not aware of the intricacies of their chosen profession. Combined with the challenges of cultural adjustment faced by all college students, these STEM-specific challenges influence many novice students in their decisions to leave STEM disciplines.

Considerable dialogue and collaboration between faculty members in the science and mathematics departments at UWG resulted in the development of a National Science Foundation (NSF) STEM Talent Expansion Program (STEP) proposal to address the retention of students in the STEM disciplines and at the institution. This project included first-year learning communities (LCs), summer research experiences, additional in-class and out-of-class support using peer-lead team learning and/or supplemental instruction, and faculty development for participating STEM faculty members. As the centerpiece of the first-year LCs, three seminar courses were developed and will be the focus of this chapter. Here, we describe three first-year seminars, each of which addresses different facets of first-year STEM student success and report on the successes and challenges this model presented for both students and faculty members.

FRAMEWORK FOR FIRST-YEAR SEMINARS

The principal goal of the first-year seminar courses was to introduce students to the contextual relevance of introductory science and mathematics courses, providing a "keystone" experience that enables students to engage in science and mathematics within the context of their professional goals. With this in mind, we designed seminars that drew on current events topics, areas that were thought to spark students' motivation for learning and enthusiasm for careers in STEM disciplines. The course What Do You Know about Health Professions? (WDKA: Health Professions) made a concerted effort to introduce students to a variety of professions tied with human health through a case study that focused on the avian flu; this seminar also introduced connections between biology and chemistry as disciplines, an increasingly important trend. In What Do You Know About Forensics? (WDKA: Forensics), students used a variety of research-grade instrumentation, problem solving, and critical thinking to unravel a murder mystery. What Do You Know About Space Science? (WDKA: Space Science) introduced students to the relevance and integration of general chemistry to a variety of STEM disciplines with special emphasis on their use in engineering related to Space Science.

Each seminar was structured to provide an intellectually exciting and academically relevant experience for entering STEM students by using active learning methods, multidisciplinary experiences, and skill development in the context of real-life problems. The students were expected to increase their own motivation levels by identifying the relevance of the course materials themselves in an active learning environment. This approach draws on lessons from the National Research Council Report (2000) that affirms "through meaningful interactions with their environment, with their teachers, and among themselves, students reorganize, redefine, and replace their initial explanations, attitudes, and abilities" and encourages development of instructional models that "seek to engage students in important scientific questions, give students opportunities to explore and create their own explanations, provide scientific explanations and help students connect these to their own ideas, and create opportunities for students to extend, apply and evaluate what they have learned."

The seminars were two-credit hour courses limited to members of three distinct LCs, with each cohort comprising 16–25 first-semester STEM freshmen. The courses met 1 day a week for 2.5 hours and were taught by an interdisciplinary team of instructors. The LC students shared a common academic schedule that included at least three other courses, including English, precalculus or calculus, and general chemistry. Because students shared the same academic schedule, they became well acquainted with one another and generally formed small groups readily. The seminar courses were laboratory- or activity-based, and teamwork and critical thinking were emphasized. Collaboration among faculty in disparate disciplines promoted the continuity of the sciences and implicitly communicated the interrelatedness and relevance of each discipline to the STEM curriculum and workplace. The LC model allowed effective communication about expectations and learning objectives and provided a shared experience across disciplinary boundaries.

WDKA: HEALTH PROFESSIONS

The unique feature of the WDKA: Health Professions seminar was the use of a current health topic, such as the avian flu, as a case study that exposed students to a potential real-world problem and familiarized them with relevant techniques used in the public health sector. Visits to health facilities confirmed the relevance of chemistry and other scientific skills in the students' future workplace and the nature of the profession.

In an initial class meeting attended by all the instructors (two chemists, a biologist, and a geographer), students were introduced to a case study pertaining to the avian flu. The case study was based on a fictitious story in which several people fell ill upon returning to the United States after visiting selected cities around the world. During a discussion, students discussed several relevant factors related to the spread of the avian flu and recognized the roles of biologists, chemists, and geographers in understanding the spread of a potential pandemic like avian flu. Furthermore, they understood that healthcare delivery is not confined to physicians, pharmacists and nurses; rather, several other disciplines such as immunology, drug design and development, and pharmacokinetics are intimately involved in understanding the origins and the progression of the disease, as well as in finding diagnostic tools and potential cures for any given disease.

These discussions served as the prelude to activities in subsequent weeks, described in detail below and summarized in Table 9.1. In addi-

TABLE 9.1 Overview of Activities in WDKA: Health Professions Seminar

Activity	Chemistry or Biology Concepts	Career Connections
Presentation of case study	Interdisciplinary project	Complex problem
Immunoassay	Volume measurement, dilution	Immunology and microbiology
Synthesis and drug design	Bonding, Lewis structures, chemical reactions, microbiology	Drug development
Drug development and pharmacokinetics	Lewis structures, mass spectra	Drug design and evaluation
GIS	Bio- and geostatistical analysis	Medical geography
Campus healthcare facility visit	Relevance of classroom curriculum in workplace	Nursing, pharmacy, primary care, and health education
Local medical center visit	Relevance of laboratory curriculum in workplace	Medical imaging, blood and urine analysis, and nursing
Career seminars	Exposure to desirable academic and nonacademic attributes	Nursing, pharmacy, and medicine
Poster presentation	Oral and written communication skills	Overview of professions

tion, students visited the campus student health facility, a local hospital and interacted with invited speakers (typically a registered nurse, a pharmacist, a physician, and an admissions officer from a pharmacy school). The course culminated in a poster session in which students, working in teams, presented posters describing one of the professions they had been introduced to during the course.

The first set of activities introduced students to *biological principles in the medical diagnosis* of avian flu. In an initial exercise, the students were exposed to fundamental concepts in biology that related to diagnosis and treatment of diseases. The reading material included fundamentals of virology, molecular biology, and immunology that briefly described viruses, viral replication, host–parasite relationships, and virus survival in the host's hostile environment (Fass 2000). The basic molecular biology reading material included DNA structure, DNA replication, amplification of DNA fragments using polymerase chain reaction (PCR), and analysis of PCR products (Mullis et al. 1986). Essentials of immunology included immune reaction, production of antibodies, and use of antibodies in diagnosis. This was followed by the classic phenolphthalein test to demonstrate the relative ease of contagious disease transmission, including flu, HIV, and STDs. In this exercise, one student was given a test tube containing diluted NaOH solution (representing the affected individual) and the rest of the students received test tubes with tap water, and students were asked to exchange the contents of these test tubes with other students. After two to three exchanges, diluted phenolphthalein solution was added to each test tube. Pink color represented the contamination. The source of contamination was traced to the original tube containing NaOH solution. The goal was to demonstrate how a single individual can spread the disease among the members of an interacting community.

Another hands-on activity applied DNA analysis to distinguish strains of virus (Saiki et al. 1988). Simulated body fluids from infected individuals in the case study were given to the students to isolate DNA. This exercise was done using a commercial PV92 DNA polymorphism kit[1] with suitable modifications. All required plasticware and chemicals such as buffers, primers, and templates were provided in this kit. The DNA analysis gave an idea of who was spreading which type of flu to whom. The ultimate proof of the spread of the virus was obtained via an immunological analysis using specific antibodies (Charlton et al. 2009). This exercise was designed to get the ultimate proof of the source of contamination utilizing specific antibodies against a protein from the human-to-human variant of the virus using another

[1] http://www.bio-rad.com/.

Drug design → chemical synthesis → *in vitro* safety/toxicity screening → activity against the pathogen → preclinical pharmacokinetics and *in vivo* safety/toxicity and efficacy screening using mice, rats, monkeys, or dogs, etc. → first phase of clinical trial (short duration of drug administration in small number of people) → second phase of clinical trial (longer duration with larger number of subject for safety/toxicity and efficacy) → third phase of clinical trial → FDA approval → drug marketing

SCHEME 9.1   Drug discovery timeline.

commercial enzyme-linked immunosorbent assay (ELISA) Immuno Explorer kit[2] containing all required solutions such as antibodies, antigens, color development solutions, and plasticware. A positive reaction confirmed that a person was suffering from the H5N1 avian flu.

In the next set of activities, students were introduced to the process of *drug design and drug discovery*. In the first activity, an introduction to the drug discovery process was made by a visiting research scientist. This was followed by two hands-on activities, one pertaining to the interpretation of mass spectra and another to drug synthesis and testing. The general chemistry concepts reinforced included isotopes, molar masses, bonding, Lewis structures, organic functional groups, and types of chemical reactions. The timeline, from drug discovery to the availability of a drug in the market, was developed using Scheme 9.1.

Selected parts of the schematic were elaborated via further discussions and simulations. The role of the Food and Drug Administration (FDA) to approve the drugs with concerns for safety, concordant with its mission to protect and advance public health, was further discussed with special emphasis on the complexity given the vast diversity of patients and how they respond to drugs, the conditions being treated, and the range of pharmaceutical products and supplements patients use. The ethical responsibilities of FDA professionals and the skills they must acquire to make crucial decisions as they weigh the information available about a drug's risk and benefit, to make decisions in the context of scientific uncertainty, and to integrate emerging information bearing on a drug's risk–benefit profile throughout the life cycle of a drug and from drug discovery to the end of its useful life were also elaborated. By taking example of pathogens such as HIV-1, the concept of targeted drug discovery was discussed, which targets different pro-

[2] Ibid.

cesses that take place in HIV-1 infections. Next, background information regarding the treatment of the influenza A virus, the virus that causes the flu, was provided from the Center for Disease Control.[3] A computer animation of the mechanism of the host virus with the antiviral agent Tamiflu was used to illustrate how Tamiflu inhibits the replication of viruses.[4]

The next set of activities utilized concepts of *chemistry* in the health professions, purposefully scheduled when the students were halfway through the first semester of general chemistry and were very comfortable with the concepts of atomic and molecular masses. Before the first activity, the general principles of mass spectrometry were introduced, followed by a discussion of the use of electrospray ionization mass spectrometry (ESI-MS) in the analysis of organic compounds that have potential use in the treatment of infectious diseases. Students visited a campus research facility that houses the state-of-the-art electrospray ionization mass spectrometer and were charged to analyze two mass spectra generated by the instrument by identifying the fragments formed. This was a good exercise that reinforced atomic and molar masses in the context of a "real-life" situation.

Next, the discussion of drug development was extended by focusing on the use of *combinatorial chemistry*, in which tens to thousands of compounds are synthesized simultaneously in a single reaction; the resulting "library" of compounds is then tested for the desired activity. Finally, promising candidates are chosen for additional testing. In a hands-on activity, students modeled this drug discovery process by preparing small libraries of potential antibacterial hydrazones from various aldehydes and hydrazines (Wolkenberg and Su 2001). The structurally related products were then screened for antibiotic activity using a nonpathogenic strain of *Escherichia coli*. A few drops of each library of hydrazones were placed in wells in each agar plate that had been streaked with *E. coli*. After incubating the plates overnight, the students observed the inhibition of the growth of *E. coli* in the wells containing the active antibiotic.

An innovative aspect of this course, which was nominally chemical and biological in emphasis, included *medical geography and epidemiology* to analyze the spread of avian flu. This activity reinforced the idea, very early in a student's career, that health-related issues are inherently complex and multidisciplinary in nature, extending beyond the STEM disciplines to the social sciences. In the spirit of John Snow, the father of

---

[3] http://www.cdc.gov/flu/protect/antiviral.
[4] http://www.pharmasquare.org/flash/Tamiflu.html#Start.

modern epidemiology, the students were encouraged to "think geographically" about health by using a geographic information system (GIS) to conduct two basic epidemiological analyses. In the first part, students used GIS to examine John Snow's cholera data by first mapping the data and visualizing the spatial pattern of deaths in relation to water pumps.[5] They evaluated the spatial relationships between cholera deaths and water pumps (Tufte 1997). In this evaluation, students computed spatial statistics or a numerical value that described a spatial (or geographic) characteristic of a set of data, a technique that has applications in both public health and epidemiology. They also calculated the mean center of the cholera deaths. Finally, they determined the point density ("hot spots"), which measured how many times a particular event (such as a cholera death) occurred in a particular area. In the second part of the geographic exercise, students used GIS to analyze the geography of avian flu, applying the techniques they learned with the cholera data to evaluate the spatial pattern of flu outbreaks, the mechanisms by which it is spreading unevenly across the world, and the changing geography of its impact on humans. Specifically, they examined the following: world map of avian flu outbreaks, geographic processes by which avian flu could spread, changing geography of human cases of avian flu, and the changing geography of human deaths as a result of avian flu.

In the culminating activity for the course, students prepared and presented a poster based on professions to which they had been introduced during the semester, *excluding* medicine, nursing, and pharmacy, the most "popular" medical careers at UWG. For each career choice, students described the role played in public health, employment opportunities in the state of Georgia, and the requirements to gain admission to programs that would prepare individuals for these professions. This exercise considerably broadened the students' perspectives on career choices related to human health, one of the main objectives of this course. It also introduced them to an important method of scientific communication very early in their careers.

WDKA: FORENSICS

The WDKA: Forensics seminar used the advantage of popular media to get students excited about science.[6] The overall objective of this seminar was to present a realistic image of the interdisciplinary nature

[5]http://www.ph.ucla.edu/epi/snow.html.
[6]http://science.nsta.org/enewsletter/2003-03/newsstories.htm.

of the STEM laboratory skills necessary to solve a crime. In addition, techniques of evidence collection at simulated crime scenes and a final mock trial allowed development and utilization of communication skills, critical thinking, and problem-solving strategies. Eight faculty members, representing chemistry, biology, physics, mathematics, computer science, and geosciences led activities in this seminar, presenting a cross section of STEM faculty to the students.

With the help of the campus chief of police on the first day, the class was introduced to a simulated crime scene and was charged to find and to convict the perpetrator(s) during the semester. The students were given a police report, an initial autopsy report, and were then taken to the crime scene and assigned specific duties. The students processed the crime scene by taking pictures, sketching the scene, identifying and marking evidence, documenting the location of evidence by triangulation, and collecting and logging the evidence. The students wrote a complete crime scene report and decided what they needed to know to solve the crime. In subsequent weeks they carried out a number of activities, summarized in Table 9.2 and described in detail below, to help them solve the crime.

The purpose of the first activity was to determine the location from which the perpetrator shot the victim. The students were given the entry angle of the bullet into the victim from the autopsy report. Each team was assigned one location and was sent out to prove or disprove that location by using trigonometry to measure the height of the landmarks. Then, using this measurement, they determined the angle of entry of the bullet into the victim and the distance the bullet traveled (Larson et al. 1993). This activity reinforced the need for accurate and precise measurement and the use of mathematical analysis in scientific investigations.

This was followed with a computer science activity where the students learned how impersonators can use the SMTP protocol on telnet sessions to send fraudulent e-mail messages.[7] They also used e-mail headers to determine the origin of the e-mail[8] and steganography software called S-Tools to test for and to reveal a hidden image embedded in a picture found as an e-mail attachment (Kessler 2004). This activity emphasized the use and abuse of communication technology, both in their personal and professional lives.

The next activity involved determining the time of death using Newton's law of cooling to model a body's loss of heat to the environ-

[7]http://www.activexperts.com/activemail/telnet/.
[8]http://www.irisonthego.com/inetSIG/2003jun2.html.

TABLE 9.2 Overview of Forensics Seminar Activities

Week	Crime-Related Activity	STEM-Related Laboratory Activities
1	First crime scene	Sampling and evidence collection (scientific method)
2	Location of the shooter	Measurements, projectiles (trigonometry)
3	Analysis of e-mail accounts	Steganography, fraudulent e-mail (computer science)
4	Time of death	Newton's law of cooling (physics)
5 and 6	Drug testing	Colorimetric testing, thin-layer chromatography, gas chromatography–mass spectrometry (chemistry)
7	Soil analysis	Scanning electron microscopy to identify mineral content of soils (geosciences)
8	Second crime scene	Sampling and evidence collection (scientific method)
8	Skid mark analysis	Coefficient of friction (physics)
9	Fingerprints and time of death	Chemical method for developing prints (chemistry) and entomology (biology)
10	Fiber analysis	Polymer identification and testing (chemistry)
11&12	DNA and blood typing	Genetics and blood types (biology)
13	Georgia Bureau of Investigation guest speaker	Career counseling
14	Mock trial	Communication, problem solving, and critical thinking

ment. The students collected time-dependent temperature data to determine the amount of heat lost by water in a coffee cup caloriometer. These data were combined with the physiological rate at which bodies cool (0.56–0.83°C/h or 1.0–1.5°F/h) to estimate the time of death of the victim (Khallaf and Williams 1991; Hasbun and Keller 1998).[9] This activity was targeted to highlight the use of physical principles in solving a biological problem.

The identification of drugs of abuse represents the most prominent use of chemistry in forensics (Berry 1985; Hasan et al. 2008). Activities were developed that focused on the techniques used by forensic chemists to identify illicit drugs, including both presumptive and confirmative tests. These activities were applied to laboratory chemicals that could be purchased without a drug license. The first activity involved

[9]http://education.ti.com/.

the presumptive identification of a powder discovered on the victim using colorimetric testing. We used commercially available narcotic identification kits (NIKs),[10] which are used by many law enforcement agencies for the identification of narcotics and drugs of abuse. In these simple tests, a color change occurs to indicate a positive test. Using the NIK Test G and diphenhydramine (Benadryl), students obtained a false positive result for cocaine. The students then did a confirmatory test using gas chromatography–mass spectrometry (GC/MS) to determine that the true identity of the drug was diphenhydramine. The next activity used thin-layer chromatography (TLC) to identify the active ingredient in pills found at the crime scene. This activity was a slight modification of the TLC laboratory done in many sophomore organic laboratories, the identification of the components of an Excedrin tablet (Williamson et al. 2007). The labels on the aspirin, caffeine, and acetaminophen knowns were changed to Ecstasy, Valium, and Oxycontin and were compared to the unknown pills. The last drug activity involved the extraction of basic drugs from the suspects' urine and analyzing the extract using GC/MS (Cochin and Daly 1963). Through these activities, freshmen students were exposed to advanced analytical techniques and instrumentation.

All the aforementioned analyses involve permanent alteration of the samples. Typically in forensic investigations, sample size is finite and some nondestructive analytical techniques need to be employed. The next activity was designed to make students aware of nondestructive analytical techniques used in scientific investigations. Using a scanning electron microscope and energy-dispersive spectroscopy (SEM-EDS), the students identified soil collected from the shooter's perch as a match to a unique soil type on campus, which also contained lead fragments (Murray and Tedrow 1975). An abandoned rifle range on campus was identified as the likely source of the admixed soil and lead. The class investigated this area of campus, looking for clues to the murderer or the murder weapon but instead found a second victim dumped in the woods in a trash bag. According to the preliminary autopsy report, this victim died of a massive trauma (hit by a car).

An important task in auto accident reconstruction involves the analysis of skid marks.[11] Armed with the length of the skid marks and the coefficient of friction, investigators can make a good estimate of a car's speed just before the driver hit the brakes. In the activity, the students determined the coefficient of friction of wood on wood using two

---

[10]http://uritoxmedicaltesting.com/narcotic-id-drug-test-kit.html.
[11]http://www.bsharp.org/physics/skidmarks.

methods typically found in an introductory physics laboratory. The students elaborated on these findings when they discussed a hypothesis that the coefficient of friction changes when the materials involved change. After doing the experiment, the students used the skid mark distance from the crime scene given to them in the police report and the coefficient of friction to calculate the car's speed in the parking lot to determine if the accident was intentional or not. This activity was aimed to make the students aware of variations in frictional resistance due to different materials.

A common piece of evidence at any crime scene is fingerprints. Fingers have a unique pattern of ridges, and along these ridges are pores that secrete biomolecules (water, oils, fatty acids, esters, salts, urea, and amino acids). Latent fingerprint are invisible prints made up of these biomolecules left on an object. The students used a variety of chemical methods (Sodhi and Kaur 1999; Nikitakis and Lymperopoulou 2008) to develop latent prints, which included dusting with powder, which relies on the mechanical adhesion of the powder to the moisture and oily components of the print; ninhydrin, which reacts with the amino acid components of sweat; and super glue fuming, in which the cyanoacrylate ester undergoes polymerization when it comes into contact with the moisture in latent fingerprints. This activity illustrated the use of chemical methods to visualize the differential distribution of biomolecules that gives rise to the unique morphological patterns that fingers leave behind.

The identification of fibers is a valuable component of many forensic investigations. In many cases, matching the fibers from the scene of the crime with those from the clothing of the suspect or from the home or car of the suspect is a critical part of the case (Deedrick 2000). Students used microscopic imaging to study the whole fiber or a cross section of the fiber sample; differential staining, where the fibers were dyed with specific stains and were identified by characteristic colors they produced; burning test, where observations were made about the way the fibers react in a flame; and solubility tests, where the fibers were tested in a range of solvents including acids, bases, and organic solvents (Fan 2005). In the activity, each group of students was provided with a reference[12] composed of a 3/8-in. wide strip of 13 different materials and the unknown fibers found at the crime scene. This activity illustrated the physical and chemical differences within and between natural and synthetic fibers/polymers.

---

[12]http://www.testfabrics.com.

Individual differences are ingrained in the DNA sequences within our genes and form the basis for combined DNA index system (CODIS). The DNA profiles in this database, based on 13 different regions in the human genome, have been extensively used in crime scene investigations. To determine the DNA profile of the victim, the suspects, and the evidence found at the crime scene, students used a kit based on one of these 13 CODIS, called "alu" polymorphism. The procedure involved isolation of DNA and amplification by PCR from saliva and blood samples at the crime scene and from victims and suspects (Mullis et al. 1986).[13] The PCR product would have a size distribution indicating one of three possible combinations—individuals with alu insertion (~300 DNA base pairs [bp]) on both homologous chromosomes would have a single 941-bp band; individuals with alu insertion on one of the homologous chromosomes would have 941- and 641-bp bands, and individuals without alu insertion would have a single 641-bp band. Students were provided additional CODIS data to validate the DNA evidence in the mock trial. In this exercise, the students were exposed to the chemical composition of DNA and bioanalytical techniques including PCR and DNA electrophoresis.

The course ended with a mock trial. Students volunteered to be lawyers and expert witnesses. The students were divided into two teams, the defense and the prosecution, with specific roles to present evidence discovered in each class activity. Each student was required to turn in a written document summarizing his or her testimony supported by research. Furthermore, they presented the evidence, developed and defended their ideas, showed an understanding of the limitations of scientific data and its applicability to law and to appreciate the importance of integrating data analysis with communication skills.

WDKA: SPACE SCIENCE

WDKA: Space Science activities were based on selected topics in matter and energy, the two overarching themes in the first semester of general chemistry. Eight faculty members from the chemistry, geosciences, mathematics, and physics departments took part in this seminar. Since all students in this course take general chemistry in the first year and calculus-based physics in the second year, the activities incorporated vocabulary from these courses. Whenever possible, faculty leading the activities consciously used scientific terms from other

---

[13] http//science.nsta.org/enewsletter/2003-03/newsstories.htm.

TABLE 9.3 Space Science Activities Related to Chemistry Concepts and Physics Topics

Chemistry Concepts	Physics Topics	Engineering Applications
Measurements and components of matter	Measurements	Designing and building a model of the ISS
Stoichiometry, gas laws, and energy transformation and conservation	Gas laws, energy transformation and conversion	Building and launching rockets with different engine types
Properties and reactions of matter	Properties of materials	Structure and properties of polymers
Thermochemistry and strength of chemical bonds	Thermodynamics and calorimetry	Fuel combustion
Redox reactions and electromagnetic spectrum	Electromagnetic spectrum, electricity	Building and testing efficiency of solar cells
Atomic spectra and spectrophotometry	Atomic spectra and spectrophotometry	Earth observations and satellite imaging

sciences, in order to illustrate connections among STEM fields. For example, stoichiometric calculations were included in activities led by physicists, Ohm's law principles guided activities led by chemists, and satellite images discussed by geologists focused on useful features of the electromagnetic (EM) spectrum. Each activity was linked to a suite of STEM concepts and was designed to highlight the connections between and among STEM disciplines. Fundamental chemical and mathematical concepts were woven through nearly every activity, introducing students to the idea that chemistry and mathematics play a central role in STEM disciplines and were relevant to all STEM fields, including engineering. In addition, activities were scheduled to align with topics being covered in lecture and laboratory in general chemistry. The fundamental chemistry concepts, physics topics, and engineering applications are detailed in Table 9.3.

While concepts in scientific measurements and properties of matter were being addressed in their general chemistry course, students participated in a 2-week activity in the WDKA: Space Science course where they designed and built a model of the International Space Station (ISS) adapted from an activity developed by the National Aeronautics and Space Administration (NASA).[14] They researched the history, design, and facilities of the ISS as well as the nature of science missions on the ISS. They were then provided with a variety of

[14] http://www.nasa.gov/pdf/136201main_International.Space.Station.pdf.

polymeric building materials and were charged to produce a functional 3-D model of the ISS that had strict mass and volume constraints. The initial part of the activity acquainted the students with the steps in an engineering design process where a word problem is first translated into a mathematical model, further refined into a 2-D drawing and finally built into a 3-D model. The steps that led to the 2-D drawing drew on the students' comprehension abilities and mathematical skills in algebra, trigonometry, and geometry. The next step involved choosing appropriate materials for the various parts of the ISS model based on mechanical properties such as strength, hardness, toughness, elasticity, plasticity, brittleness, ductility, and malleability. They were then required to further modify and refine the materials list based on density measurements to meet the mass constraints of the model. Finally, students measured appropriate quantities of materials and assembled them into the final product while realizing the relevance of accurate and precise measurements to control the cost of the product. To culminate the activity, each team presented the model to the class with special emphasis on the functionality and safety of their design.

Most engineering students treat stoichiometry as a mere mathematical exercise and fail to see how the chemical basis of the topic is relevant to various energy-related issues in the engineering world. During the 4 weeks when the general chemistry course focused on topics of stoichiometry, gas laws, and thermochemistry, the WDKA: Space Science LC students studied the chemical and thermal aspects of two combustion reactions that showed how fundamental chemistry is used constructively in engineering and in physics. In the first combustion activity, students built a variety of model rockets from easy-to-assemble kits, available from Estes.[15] On launch day, rockets with a variety of engine types (coded depending on their total impulse and average thrust) were launched, recovered, reloaded, and relaunched to make comparative observations and measurements about the flight profile/ maximum altitude and thrust generated. Differences in engine performance depended on the structure, properties, and amounts of the selected propellant as well as the nozzle design (excluded from further discussions due to limited fluid mechanics knowledge of the students), which provided the necessary thrust when high-temperature and high-pressure product gases escaped the combustion chamber within the engine in accordance with the gas laws. Concepts in chemical structure, properties, and stoichiometry provided a basis for understanding fuel

---

[15] http://www.esteseducator.com/Pdf_files/Tech%20Manual.pdf.

quantities and engine performance. According to the law of conservation of energy, the chemical energy was converted to mechanical energy and resulted in the rockets being launched to reach altitudes that were proportional to the amount of propellant present in the various engines.

In the second combustion activity, students studied fuel combustion by comparing the burning characteristics and heating values of a variety of liquid fuels in a calorimeter (Rettich et al. 1988). Using a coffee cup calorimeter, a measured volume of water was heated by the combustion reaction of fuels like ethanol, vegetable oil, gasoline, diesel, and motor oil. The calorimeter was calibrated by burning a fuel, such as methanol, with a known heat of combustion. Using basic principles of calorimetry, the heat released by the fuel was a result of combustion and gained by the water. This quantity could be calculated easily from its mass, temperature change, and specific heat capacity. The heating value of the fuel was measured by the heat released per unit mass of fuel burned. The effect of reaction stoichiometry on the heat output and carbon efficiency was also studied by varying the amounts of fuel. Visual inspection during the burning, such as the presence of soot, allowed evaluation of pollution effects. Analysis of experimental data initiated conversations about the chemical composition of the fuels, the composition of gasoline, and the meaning of "octane number." Experimental data were compared to empirical thermodynamic equations such as the Dulong formula that are commonly used in engineering calculations (Lloyd and Davenport 1980). Both these activities provided valuable discussions about the economic and environmental impact of modern fuel technology and the need to explore alternate energy sources.

Appreciating the interdependence of chemical structure and bonding in general chemistry helps selection of functional materials for use in engineering applications. Polymers are one of the most versatile materials used in the engineering world and have been studied using some concepts included in Macrogalleria.[16] Initial discussions revealed the different types of polymers that make up modern cars including acrylonitrile–butadiene–styrene body parts, Kevlar tires, polyisoprene wipers, polycarbonate headlight lenses, cellulose air filters, polyvinylchloride pipes, and nylon carpets. Applications of polymers in the construction industry, electronics, pharmaceuticals, sporting goods, clothing, and other consumer products were discussed to show the many unique properties of polymeric materials that can be attributed to their chemical structure. The differences between the polymers,

[16] http://pslc.ws/macrog/maindir.htm.

attributed to three factors including chain entanglement, intermolecular forces, and time scale of motion, were studied using polyvinyl alcohol (PVA) solutions used for making slime using simple viscometers. In addition, the water-absorbing ability of polyacrylates in diapers and the effects of a variety of salts on the gel were explored (Cleary 1986; Criswell 2006). Two other types of polymerization were carried out involving the cross-linking and de-cross-linking of alginates in the presence of calcium and sodium ions as well as the formation of nylon and caprolactum (Friedli and Schlager 2005).

Interaction of EM radiation with matter in the form of absorption, emission, and scattering leads to the useful field of spectroscopy. Although the use of spectroscopic instrumentation is mostly in the realm of scientific research laboratories, the engineering practice often uses spectroscopic principles in various applications such as solar cells and satellite imaging. In the first EM activity, students were exposed to solar energy and to silicon-based and nanocrystalline solar cell technologies. Using a prototype photovoltaic cell kit[17] and natural dyes extracted from berries, students learned the basic principles of biological extraction, chemistry, physics, environmental science, and electron transfer and applied them to an engineering problem, specifically the problem of building a functional solar cell. The role of specialty materials in the efficiency of solar cells such as conductive glass slides, nanocrystalline titanium dioxide, and an iodide electrolyte was studied by varying these materials during the building phase.[18] The redox reaction between photons from solar radiation and molecules of the dye was facilitated in the presence of the nanocrystalline particles, and the electrolyte helped in the regeneration of the dye. The study on the photogeneration of electricity using solar cells utilized principles of circuit design.

In the second EM activity, students were introduced to the technique of collecting EM information about the earth from afar, either from airplanes or from satellites, via remote sensing that involves the joint efforts of numerous engineers, technologists, and scientists who provide us a global view of our planet. Students used handheld spectrometers, a type of active remote sensing instrument, which provide their own energy in the form of EM radiation to illuminate the object or scene they observe and send a pulse of monochromatic light from the sensor to the object and then measure the radiation that is reflected or backscattered from that object. In this way, students explored the basic method to

---

[17] http://ice.chem.wisc.edu/.
[18] http://www.solideas.com/papers/Exploratorium_Solar.pdf.

detect, measure, and analyze the spectral content of EM radiation reflected from diverse materials. Students connected the "theoretical" concepts of color and spectroscopy to the tangible and applied outcome of generating a satellite image based on reflectance properties of the earth's surface. Students extended this thread by researching topics of their choice, in global and environmental change. In this third EM activity, students evaluated images acquired by LANDSAT and other satellites, which use passive remote sensing methods, in which the satellite detectors use diffraction gratings and/or prisms to discriminate among wavelengths of sunlight reflected from the earth's surface and to create maps of surface phenomena.[19] At the end of the activity, students presented their findings, including the use of remote sensing technology, as a tool in evaluating global-scale change phenomena.

RESULTS

Overall, the interdisciplinary WDKA seminars have been successful in the five years since their introduction at UWG. Both student satisfaction and student success data indicate that the seminar courses made a positive impact. Further, we found that the level of interaction among faculty was high and produced positive collaborations. Finally, the courses posed logistical challenges that may be difficult to overcome in the absence of external funding or tangible institutional support.

Course evaluations indicated that students enjoyed the courses overall and were beginning to understand the relevance of biology, chemistry, physics, geosciences, computer science, and mathematics to their STEM careers. Students particularly appreciated the inquiry-based, hands-on, and applied nature of the activities, reporting that they found general chemistry, biology, physics, and geosciences more interesting, relevant, and, easier to understand. They enjoyed the variety of instructors and the passions they brought about their disciplines into the classroom. Evaluations also showed that student perception of integration among disciplines improved significantly between the first and second years of each LC cohort.

One of the main goals of the LC seminar courses was to increase first-year success in key science and mathematics courses. Specifically, we targeted success in general chemistry and introductory mathematics courses. All general chemistry students take the American Chemical Society (ACS) general chemistry exam at the end of each term. Students

---

[19]http://earthobservatory.nasa.gov/Features/RemoteSensing/remote.php.

TABLE 9.4   Comparison of Introductory Mathematics Success Rates[a] during the First Year

Cohort	Fall 2005 (%)	Fall 2006 (%)	Fall 2007 (%)	Fall 2008 (%)
Health professions	—	86	84	87
Health—comparison[b]	—	30	29	53
Forensics	75	54	74	56
Forensics—comparison[b]	27	34	32	34
Space	61	88	83	81
Space—comparison[b]	45	57	48	48

[a]Success rates represent the percentage of students successfully (C or higher) completing precalculus or higher math course during their first year.
[b]The cohorts for comparison were chosen by a procedure established at UWG's Institutional Research Office to produce a set of non-LC students with similar entering characteristics (GPA, test scores, majors) to minimize program selection bias.

in the three LCs performed better on this exam at the end of general chemistry than did non-LC students; on average, the score for LC students is 39, compared to a departmental average score of 35. The combination of additional science content, deliberate integration of chemistry with other science disciplines, and increased appreciation of chemistry's relevance to students' career paths likely contributed to this success.

Additionally, LC students are more likely to successfully complete at least one semester of introductory mathematics for science majors (precalculus or higher) during their first year than are comparable science majors not in one of the LCs (Table 9.4). In this case, students in the LCs benefit from some structural aspects of the community; all students in the LCs are placed in precalculus or calculus during their first semester. Students in the comparison group, despite being eligible to take precalculus or calculus, frequently do not attempt either course, opting for college algebra instead. Additionally, the first-year seminars deliberately integrate mathematical content with the intention of highlighting its relevance to the students. This integration has the additional effect of encouraging students to complete necessary mathematics courses.

Faculty members were pleased with the level of enthusiasm and engagement on the part of the students and saw a deeper understanding of scientific concepts beyond the mathematical equations. Through engaged inquiry learning within a collaborative and cooperative environment, LC students developed higher-level cognition as they mastered content and process skills. According to traditional measures of first-year success—first-year GPA and first-to-second year retention rate—these LC students were generally more successful compared with students without such a framework (Table 9.5). Although the students in first-semester seminars were usually successful, individual cohorts varied from year to year, and success by these measures is not a foregone conclusion. For example, both the first-year GPA and retention rate in the WDYK: Forensics cohort were lower than those of the comparison group. However, the success rate in introductory mathematics courses was consistently higher, suggesting that many non-LC beginning science students did not attempt precalculus or calculus in their first year, thus putting them behind in progress toward a science major. In addition, students in these communities are retained as STEM majors more frequently than average freshmen and are accepted as transfer students (for engineering majors) at a higher rate (50% for LC students vs. 35% for non-LC engineering students).

TABLE 9.5 Student Success Measures for Students in WDKA Courses

Cohort[a]	Retention (%)[b]	Year 1 GPA[c]
2004 WDYK: Forensics	76	2.74
Comparison Group	76	2.72
2005 WDYK: Forensics	67	2.16
Comparison Group	74	2.69
2005 WDYK: Space Science	78	2.39
Comparison Group	59	2.33
2006 WDYK: Forensics	92	2.74
Comparison Group	75	2.67
2006 WDYK: Space Science	83	2.45
Comparison Group	79	1.61
2007 WDYK: Forensics	74	3.10
Comparison Group	80	3.00
2007 WDYK: Space Science	79	2.81
Comparison Group	70	2.61
2007 WDYK: Health Professions	76	2.92
Comparison Group	75	2.51
2008 WDYK: Forensics	Not available	2.63
Comparison Group	—	2.84
2008 WDYK: Space Science	Not available	2.65
Comparison Group	—	2.55
2008 WDYK: Health Professions	Not available	2.99
Comparison Group	—	2.39

[a]The cohorts for comparison were chosen by a procedure established at UWG's Institutional Research Office to produce a set of non-LC students with similar entering characteristics (GPA, test scores, majors) to minimize program selection bias.
[b]Retention measured as fall-to-fall enrollment from first to second year.
[c]GPA measured at the end of the first spring semester.

Extending beyond traditional introductory content, the activities in these seminar courses were integrative and showed how interconnected topics in biology, chemistry, physics, and mathematics could be used to address complex problems in science. Team teaching allowed content to be covered in appropriate technical depth while keeping the focus and emphasis of the course at a level that interests and engages all students. In addition, a team faculty approach provided an opportunity to include social, economic, cultural, and global perspectives on science and engineering and brought diverse perspectives to bear in a freshman course. Faculty expressed great enthusiasm for working across disciplines on this project, reporting that they better understood what kinds of pedagogy and content occurred in other departments. They also enjoyed the challenge of integrating their own discipline within the established themes of each seminar. Several faculty members participated in more than one seminar, and one participated in all three.

These interactions among faculty also facilitated transformations in introductory science and mathematics course instruction. These course modifications were an integral part of the STEP grant, but the high level of cooperation among faculty was made possible in part by the fact that faculty were teaching the same students simultaneously and were engaged in conversation about teaching. In particular, mathematics, biology, and physics adopted a supplemental instruction model similar to the peer-led team learning model already implemented in general chemistry courses. Increased attention to student success and student challenges, coupled with specific institutional support for math tutoring, resulted in overall improvement of success rates in entry-level mathematics courses.

Classroom interactions among faculty also stimulated research collaborations. Several interdepartmental proposals and projects have been launched among the faculty who participated in the LCs. For example, six of the project faculty members are involved in a privately funded multiyear outreach program. Several of the activities implemented in this outreach program (for K-8 students) are modified from activities done with the LC freshmen. In addition, research proposals to federal agencies have emerged between faculty members involved in this project. The enhancement of interdepartmental synergy was a positive and unanticipated outcome of faculty team teaching in these courses.

These courses, while successful at many levels, pose a set of challenges not uncommon in collaborative learning environments, namely, that the courses are time intensive, personnel intensive, and require careful planning and coordination to maximize learning for the students. Each course had a designated coordinator, who worked with the first-year program office to recruit and enroll students and to schedule classes. These coordinators were responsible for recruiting instructors to assist in the course and for doing the overall course planning. Each course additionally had three to six additional instructors that directed one or more course meetings. The first-year seminar courses were most successful when faculty communicated and collaborated effectively, producing a course that appeared seamlessly integrated from the students' perspective. For the duration of the STEM grant, each participating faculty member was compensated for each course meeting. The coordinator was additionally compensated by the first-year program office for directing the LC. Although participating faculty report that compensation was not the main motivation, most believed it to be important, particularly when participation in such a course does not "count" toward calculations of total teaching hours. It is critical to have an institutional structure that recognizes the importance of this kind of teaching contribution, either by compensation or by acknowledgment in a workload policy. In the absence of such support structures, courses like these are difficult to maintain over the long term, even when they are successful.

CONCLUDING REMARKS

All three seminars were targeted to address (1) overwhelming content by providing a collaborative/cooperative learning environment; (2) connections among content areas by building interdisciplinary, inquiry-oriented activities; and (3) content relevance by providing a model for contextualization of "theoretical" mathematics and science content.

The course WDYK: Health Professions focused on the avian flu, an important news item when this course was designed. It can easily be modified to incorporate swine flu, which is of current interest. Interestingly, instructors who use *The New York Times* or the National Public Radio in their courses can tie current or archived materials on any virus-related ailment in these media to the content of this course.

Similarly, WDYK: Forensics made use of the popularity of the *CSI* television series to draw science students into interdisciplinary scientific learning. The application of a wide variety of research-grade instrumentation to problem solving, combined with a mock trial, gives students the opportunity both to generate relevant data and to critically evaluate the security of their conclusions. Inquiry-oriented science seminars such as this could focus on a wide variety of applications in addition to crime scene analysis, including environmental contamination, risk evaluation, or epidemic analysis.

The integrative laboratories in WDKA: Space Science showed how fundamental topics in chemistry, physics, and mathematics can be interconnected to address complex problems in engineering. They also provided an opportunity to include social, economic, cultural, and global perspectives along with science and engineering.

The connections that students in all these courses made to a wide array of STEM careers are obvious. An important underlying theme, albeit a subliminal one, is the students' exposure to state-of-the-art instrumentation along the way, particularly in the pre-health and forensics LCs. These included GC/MS, ESI-MS, and scanning electron microscopy (SEM) ubiquitous in modern laboratories that focus on environmental and forensic analyses, as well as research laboratories in biology and chemistry. Over the past five summers, several students in the three LCs carried out faculty-directed research for one or two months over the summer following their first year or during their sophomore year under the auspices of the STEP grant. In many instances, their research projects used these instruments, as well as UV–vis, infrared (IR), and nuclear magnetic resonance (NMR) spectrometers. Most of these students will pursue an undergraduate degree in chemistry, biology, engineering, or geology. Regardless of their career choices, it can be safely said that these seminar courses helped initiate or sustain the students' interest in the sciences.

ACKNOWLEDGMENTS

The seminars were funded by the NSF STEP grant DUE-0336571. Any opinions, findings, and conclusions or recommendations expressed in this material are those of the authors and do not necessarily reflect the views of the NSF.

The authors are thankful to Ghazia Asif, Ben de Mayo, Rebecca Dodge, James Espinosa, Javier Hasbun, Michelle Joyner, Deborah Lea-Fox, Dusty Otwell, Lori Payne, Nancy Pencoe, Muhammad Rahman, Karen Smith, and N. Andy Walter for their help in developing, implementing, and teaching the seminar courses.

REFERENCES

Berry, K. O. (1985) *J. Chem. Educ. 62*, 1044–1046.

Charlton, B., B. Crossley, and S. Hietala (2009) *Comp. Immunol. Microbiol. Infect. Dis. 32*, 341–350.

Cleary, J. (1986) *J. Chem. Educ. 63*, 422–423.

Cochin, J. and J. W. Daly (1963) *J. Pharmacol. Exp. Ther. 139*, 154–159.

Criswell, B. (2006) *J. Chem. Educ. 83*, 574–576.

Deedrick, D. (2000) *Forensic Sci. Commun. 2*(3), http://www.fbi.gov/hq/lab/fsc/backissu/jan2004/research/2004_01_research01b.htm (accessed September 29, 2009).

Fan, Q., ed. (2005) *Chemical Testing of Textiles*. Boca Raton, FL: CRC Press.

Fass, M. F. (2000) *Microbiol. Edu. 1*, 20–25.

Friedli, A. C. and I. R. Schlager (2005) *J. Chem. Educ. 82*, 1017.

Hasan, S., D. Bromfield-Lee, M. T. Oliver-Hoyo, and J. A. Cintron-Maldonado (2008) *J. Chem. Educ. 85*, 813–816.

Hasbun, J. E. and G. E. Keller (1998) http://www.home.agilent.com/upload/cmc_upload/All/exp75.pdf (accessed September 29, 2009).

Kessler, G. C. (2004) *Forensic Sci. Commun. 6*(1), http://www.fbi.gov/hq/lab/fsc/backissu/jan2004/2004_01_book01.htm (accessed September 29, 2009).

Khallaf, A. and R. W. Williams (1991) *J. Forensic Sci. Soc. 31*(1), 7–19.

Larson, R. E., R. P. Hostetler, and B. H. Edwards (1993) *Precalculus: A Graphing Approach*. Lexington, MA: D. C. Heath and Company.

Lloyd, W.G. and D. A. Davenport (1980) *J. Chem. Educ. 57*, 56–60.

Mullis, K., F. Faloona, S. Scharf, R. Saiki, G. Horn, and H. Erlich (1986) *Quant. Biol. 51*, 263–273.

Murray, R. and J. Tedrow (1975) *Forensic Geology: Earth Sciences and Criminal Investigation*. New Brunswick, NJ: Rutgers University Press.

National Research Council (2000) *Inquiry and the National Science Education Standards: A Guide for Teaching and Learning*. Washington, DC: National Academy Press.

Nikitakis, A. and K. A. Lymperopoulou (2008) *J. Chem. Educ. 85*, 816A.

Rettich, T. R., R. Battino, and D. J. Karl (1988) *J. Chem. Educ. 65*, 554–555.

Saiki, R. K., D. H. Gelfand, S. Stoffel, S. J. Scharf, R. Higuchi, G. T. Horn, K. B. Mullis, and H. A. Erlich (1988) *Science 239*, 487–491.

Sodhi, G. S. and J. Kaur (1999) *J. Chem. Educ. 76*, 488A.

Tufte, E. (1997) *Visual Explanations: Images and Quantities, Evidence and Narrative*. Cheshire, CT: Graphics Press.

Williamson, K., R. Minard, and K. Masters (2007) *Macroscale and Microscale Organic Experiments*, 5th ed., pp. 183–185. Boston: Houghton Mifflin.

Wolkenberg, S. E. and A. I. Su (2001) *J. Chem. Educ. 78*, 784.

CHAPTER 10

The Center for Authentic Science Practice in Education: Integrating Science Research into the Undergraduate Laboratory Curriculum

CIANÁN B. RUSSELL,[1] ANNE K. BENTLEY,[2] DONALD J. WINK,[3] and
GABRIELA C. WEAVER[4]

[1]Georgia Institute of Technology, Atlanta, GA
[2]Lewis & Clark College, Portland, OR
[3]University of Illinois at Chicago, Chicago, IL
[4]Purdue University, West Lafayette, IN

Laboratory experiences have long been recognized as an important component of an undergraduate chemistry education (Armstrong 1903). Laboratory instruction became commonplace in colleges and universities beginning in the early 1900s, and today, the American Chemical Society (ACS) Guidelines on Undergraduate Professional Education in Chemistry require that programs include 400 hours of laboratory instruction beyond introductory chemistry to grant an accredited chemistry degree (ACS 2007).

Despite widespread acceptance of the importance of laboratory instruction, little consensus exists among educators about the learning goals for the chemistry laboratory and the type of laboratory curriculum that can best achieve those goals. Educators report a wide range of learning goals for the students in their laboratory courses, including developing interest in the subject, increasing motivation, learning

Any opinions, findings, and conclusions or recommendations expressed in this material are those of the author(s) and do not necessarily reflect the views of the National Science Foundation.

scientific concepts, improving laboratory skills, and developing problem-solving abilities (Hegarty-Hazel 1990; Tamir 1990; Abraham et al.
1997; Trumper 2003; Hofstein and Lunetta 2004). The laboratory is not
usually described as having specific content learning goals, but rather
goals that are largely either affective (e.g., interest and motivation) or
hands-on (e.g., practical skills).

In the most common format utilized at U.S. colleges and universities,
laboratory experiments, intended to be closely correlated with lecture
topics, provide students step-by-step instructions to gather data used to
verify factual information (Abraham et al. 1997). However, the traditional format does not give students an accurate impression of the
nature of science (Hodson 1993, 1996). Instead of developing skills used
to design experiments and to gain new knowledge, students focus on
following the instructions, completing the activity, and finding the
"right" answer in the end (Boud et al. 1980; Russell and Weaver 2008).
Students engaged in these types of laboratory activities do not feel ownership of their work and are not exposed to the scientific process.
Students experiencing traditional laboratories also do not describe their
laboratory activities as relevant to their everyday lives (Russell 2008).

A new alternative to the traditional laboratory is the research-
based laboratory, which integrates research into the undergraduate
laboratory curriculum. Research-based laboratory curricula involve
students in current research projects and give students a stake in
their work by having students collect data that will be used by scientific researchers (Chen et al. 2005; Hanauer et al. 2006; Weaver et
al. 2006). Numerous studies have indicated the positive effects of
research experiences on students in terms of retention in the sciences
and attitudes about science (Nagda et al. 1998; Wenzel 2003; Lopatto
2004; Seymour et al. 2004). Research-based laboratory curricula integrate aspects of research experiences such as decision making and
responsibility to the scientific community into students' early science
laboratories. Students know that the decisions they make in planning
their experiments will affect the data they collect and that the data
will be useful to a research group. Their work is part of a larger
context, not a redundant activity that has been completed by many
students before them (Russell 2008).

THE CENTER FOR AUTHENTIC SCIENCE PRACTICE IN EDUCATION (CASPiE) CURRICULUM STRUCTURE

The CASPiE curriculum model incorporates 6- to 8-week research
modules authored by active research faculty (CASPiE 2005). The

faculty member is supported by a development team typically consisting of the CASPiE principal investigator, a community college faculty member, a graduate student and/or postdoctoral researcher, and one or more undergraduate researchers. The development team serves to provide pedagogical assistance in the module writing process as well as perspective regarding the level of the writing, content, and tasks included in the module. Each module presents a research project that teaches chemistry concepts appropriate to either the general or the organic chemistry curriculum. Students work in teams of three in the laboratory setting and spend the first few weeks of the module learning synthetic or analytical techniques that they apply in the next few weeks to answer an authentic research question of their choosing. Beyond teaching content knowledge, the modules are designed to require students to develop and test their own hypotheses.

In a traditional research group, students interact with peers who are at various stages in their educational careers. To mimic this environment in the laboratory setting and outside of it, CASPiE students are guided by peer leaders, students who have previously completed a CASPiE module. The peer leaders emphasize the context of the project within the broader research area. The peer groups help the students grasp concepts and research skills that are usually absorbed from a mentor in a research group, such as searching the literature, extracting relevant information from research articles, designing quality experiments, and preparing presentations about their work.

To ensure that students generate reliable and publishable data, CASPiE has developed a network of instruments that can be accessed and controlled remotely over the Internet. If the necessary instruments are not available on their own campuses, students mail their samples to the instrument site and can then log in to set the parameters and run the experiment themselves. Data are continually updated by students at participating institutions via a database and internet communication.

The CASPiE curriculum is used by many institutions, including research institutions, liberal arts institutions, and community colleges on two continents. By design, the modules are constructed to work well at any type of institution and with students of diverse academic backgrounds. Participating institutions serve varying numbers of students from a broad spectrum of socioeconomic and racial backgrounds, giving large numbers of students access to the CASPiE curriculum and thus access to research; many of these students would otherwise have difficulty gaining such access independently. Because of the multi-institutional impact of the curriculum, CASPiE provides many more students with access to understanding the process and mechanisms of research than typical undergraduate research experiences can. The interplay of research and student laboratory experience will be discussed in this chapter.

AN EXAMPLE CASPiE MODULE

John R. Burgess developed a module titled, "Phytochemical Antioxidants with Potential Health Benefits in Foods," which focuses on evaluating the antioxidant capacity of foods and the impact of factors such as storage, cooking, and digestion on these antioxidants. The module is designed for the second semester of general chemistry and specifically focuses on the concepts of dilution, solution making, kinetics, and the Beer–Lambert law for UV–vis spectroscopy, which are concepts typically addressed in general chemistry and in the general chemistry laboratory.

During the early weeks of the module, students learn three methods for measuring the antioxidant capacity of samples: the Trolox equivalence antioxidant capacity assay, the polyphenolics assay, and ascorbate measurement by high-performance liquid chromatography (HPLC). Students also learn how to perform accurate dilutions and serial dilutions. Beginning in the fourth week, students have the opportunity to generate their own research project within the topic of the antioxidant capacity of foods. The students are encouraged to use the techniques that they have learned during the skill-building weeks of the module in their project development. Students then have between 2 and 4 weeks to explore their research question, to analyze their data, and to develop conclusions. Students report their data using an online form, and the Burgess laboratory uses those data to investigate current problems and to explore possible new research directions.

MODULE AUTHORS WORK WITHIN
THE DUAL NATURE OF CASPiE

CASPiE modules typically begin with 3 weeks of introductory laboratory activities followed by 3–4 weeks of student-designed research projects. The introductory laboratory activities teach the students the skills and content knowledge necessary to understand the module's research area. After completing the introductory activities, the students are prepared for the second stage of the module in which they participate in the research process. Each module has two simultaneous yet distinct goals: teaching both chemical content and the research process. Module authors face the challenge of developing a module that balances these two aspects of the students' learning throughout the experience without sacrificing one or the other.

Specific learning goals will not be achieved unless the students understand the significance of the research topic and are motivated to apply themselves to the work. In many cases, authors provide this motivation by relating the chemistry in their module to an issue already of concern to students. One author said, "Especially if someone's interested in health food, you know, and they may be buying health food, you know, they can actually see why, what does this mean, and how is it that understanding basic chemistry can allow us to understand what an antioxidant does" (Module author 1). Module authors are enthusiastic about sharing their research interests with the students, and they recognize the importance of identifying a research problem that students will be interested in solving.

The modules encourage students to realize that chemistry and chemical concepts are not limited to classroom learning; by participating in a CASPiE module, students begin to recognize the chemistry in other disciplines and the world around them. Many module authors have been recruited from outside traditional chemistry departments. One module author highlighted the importance of interdisciplinary work:

> They end up taking a bunch of, ah, classes … they rarely see the integration of knowledge. I have that come back to me with a lot of graduate students [taking] courses, "I never realized that there was this connection." I think we need to do more of it, where we integrate disciplines into chemistry because chemistry is the foundation for really understanding biology. (Module author 2)

Unlike most beginning college students, research faculty can see the connections across disciplines and highlight the role chemistry plays in their own work.

Within the broader context of their chosen research project, module authors communicate chemical concepts and introduce students to the research process. Faculty are motivated to write CASPiE modules in part because they recognize the opportunity to teach students the chemistry concepts involved in their own research. One module author described her motivation by saying, "I wanted to come up with a very simple and, and very um, elegant experiment that can explain many concepts that are the key concepts of solid state chemistry" (Module author 3). Another author, in describing the approach he took to develop the introductory activities, said, "I was thinking, are those techniques going to be tools that they could [use to] learn the concepts in chemistry?" (Module author 1). The module development team collaborates with the authors throughout the writing process to ensure that the chemical concepts within the research are emphasized and to establish connections for the students between chemical knowledge and practical applications.

The CASPiE modules that each author creates should guide students through an independent research experience in which they learn the research process by experiencing both the challenges and rewards of research. The research question each author chooses to address, therefore, needs to be, in the words of one author, "a question which is both answerable and a question which is worth answering" (Module author 4). Students need to make mistakes and then correct their mistakes. In doing so, they are making authentic contributions to the research project. One author reflected on the students' research work in his module:

> They're examining reagents which have not, have not previously been used for reduction of the solid phase and we're sort of in limbo at the moment with our new set of substrates that we're looking at but the initial results are looking promising, so the students are actually doing genuine, novel research. (Module author 4)

Authors design the modules so that students are at the edge of scientific knowledge. By keeping the modules current, module authors provide the students with an authentic research experience.

COMMUNICATING THE PROJECT CONNECTION

The CASPiE program depends on more than content knowledge from the instructors involved in teaching the course—it is important that instructors understand and acknowledge the goals of the program for those goals to be met. In particular, students enrolled in CASPiE courses are contributing to the research of a scientist: a real human being with a laboratory at a university. This idea can be difficult for students to understand. More surprisingly, instructors (be they peer leaders, teaching assistants, or faculty) also fail to grasp the concept of student contribution to real research. However, once students believe that their data are relevant to the work of a scientist, the impacts of the laboratory experience for the students are resoundingly positive.

It is certainly a problem—for the students, for the instructors themselves, and for the researcher—when an instructor does not understand that the research data their students collect are valuable to a scientist. The students can fail to make personal connections with their work, perceiving the activities as simply more of the same traditional laboratories that they have always done. Instructors can fail to understand why particular choices were made in the curriculum and can suggest changes that completely remove the research component from the laboratory. In cases such as these, the complexities of creating an environment that synthesizes research and the undergraduate teaching laboratory are evermore clear.

In one instance, a faculty instructor described his vision of changes that would improve the CASPiE curriculum in this way:

> Deemphasize [the complicated instruments and complicated molecules] and you sort of emphasize the idea of background material and what have people done and then ask them to argue, make that a larger part of the program, this hypothesis stuff, let them hypothesize more than once … and maybe if you picked a little bit simpler labs, then maybe they could actually participate more and more often. (Faculty instructor 1)

The comments from this instructor show an interest in student participation in hypothesis development, but there is a clear lack of understanding of the utility of the results. The emphasis that this instructor places on simplifying the laboratories, taking the focus away from complicated molecules, and refocusing on previous work highlights the instructor's lack of understanding with regard to the contribution that the students are making to the larger body of research data.

However, when students were able to see that their research was contributing to the work of a research scientist, they were very excited. One student described his experience in this way:

> I mean as far as, as I was, as I was concerned, it was an ordinary lab in a sense but in another sense it was more than that. There was actually essentially a research paper like other people may be using this research so, I think that's pretty cool, actually … Um, I think it would be pretty awesome because if some scientist in a couple of years were to find a revolutionary procedure for something or a revolutionary discovery and to think that he or she may have possibly used my data, I feel that I've contributed to his or her success. So, it feels like I have a piece of the success already and, ah, I might not know about it until a couple years down the line. (Student 1)

This student was very excited about the prospect of his data being used by a researcher. In expressing this excitement, the student shows that he understands not only that his work is important in the context of his course, but also that it adds to the body of knowledge in the research field. The student's feeling of contribution can be attributed to the comingling of laboratory activities and research activities that CASPiE provides.

TEACHING ASSISTANTS WILL MAKE OR BREAK CASPiE

The CASPiE curriculum depends heavily on the active participation and buy-in of personnel at all levels. However, simply supporting the pedagogical concept is not sufficient. It is important for all personnel to have clearly defined roles and expectations. The teaching assistants are precariously positioned because of the demands of their job: CASPiE teaching assistants must serve as both educators and research mentors.

Previous teaching experience generally serves to prepare teaching assistants for the role of educator in the laboratory, or teaching assistant training at their institution can provide them with this information. If by no other means, teaching assistants can learn about being an educator from their own experiences as students. However, serving as a research mentor is a wholly different role for which teaching assistants are generally underprepared. The lack of preparation primarily manifests itself in two ways: (1) teaching assistants fail to adequately respond to student questions about research and (2) teaching assistants overdirect student research.

Inadequate Teaching Assistant Response

Research is complicated. For example, there is no clear, predetermined path from point A to point B; it is rare that a procedure succeeds on the first try for a variety of reasons; it cannot be determined with absolute certainty if the answer obtained is the *correct* one. It is thus difficult for graduate teaching assistants inexperienced in research themselves to be comfortable leading students in research endeavors. Teaching assistants, especially teaching assistants early in their graduate careers, struggle with being a teaching assistant in CASPiE, as it pushes them to take on an advisory role rather than simply a dictatorial one. One teaching assistant revealed how difficult it was for her to fill the advisory role, saying,

[I] struggled with helping st-, helping students deal with, um, experimental data that they couldn't explain. That was hard sometimes, 'cause I didn't know, and so they would sort of look to me to answer something, and I think it was good for them to see that I didn't have the answer ... I think that was good for them to see that that was sometimes hard for me, 'cause I didn't wanna lose my credibility as a knowledgeable person in chemistry [laughs] but, um, but I, I didn't know what was going on in that, in that particular case. (Teaching assistant 1)

This teaching assistant describes the struggle of their role, noting the importance of students understanding that even the teaching assistants do not have all of the answers. It is likely that students have developed expectations from their previous experiences for the roles of their teaching assistants that include providing the answers to student questions. This is a central issue for the teaching assistant to understand; however, this teaching assistant goes on to explain discomfort with possibly losing credibility, which shows a deeper level of the struggle in the role of research advisor/teaching assistant that CASPiE creates. The teaching assistant struggles with not wanting students to believe that she lacks the knowledge to lead them, thus failing to make the connection between her role and that of a research mentor—one who need not know the answers to all questions.

Overdirection

One of the major tenets of the CASPiE curriculum is that students be directly involved in authentic research, including the procedures by which they address the research questions. The intention of the curriculum is that students will choose their research question within the framework of the module and design a protocol using whichever components they deem appropriate from those that they have learned. Some students may choose components that do not "make sense" (i.e., conducting an assay for a compound that is not expected to be present in their sample). The design of the curriculum intends that students will make their own decisions, whether they be "right" or "wrong." This element is purposeful: the CASPiE curriculum development team expects students to learn from their frustrations (in the example given, assays that find zero concentration because the compound is not present). When students are not given the opportunity to develop their protocols, this element is diminished and is sometimes lost.

At some institutions during the Spring 2008 semester, the implementation left the teaching assistants without enough instruction, developing into situations in which students described incidents of overdirection. In one such example, a student described her research experience by explaining that "our TA told us to just repeat the first three laboratories, just change everything to tea, so we did, so, that's how we did it" (Student 2). This student describes a situation in which the teaching assistant has led the student's research experiment so much that the students have little say in the experiment at all. The student did not experience authentic science practice because of the overdirection of the teaching assistant, but instead experienced something more akin to structured inquiry. A teaching assistant at the same institution described his perspective of the CASPiE curriculum in this way:

> One of the things they say is we're not supposed to, like, we're supposed to let them do their own thing, their own research ... You come in, you do pretty much a cookbook and I don't see much of a difference between CASPiE and the, and the cookbook labs. (Teaching assistant 2)

The statements of this teaching assistant intricately show the challenges that arise when intermingling the two identities of graduate students—teaching assistant and researcher. This teaching assistant fails to recognize the research elements in his students' work. He does not provide his students with the necessary open-ended instruction to allow them to think for themselves, but instead overdirects their projects by interacting with them in the same way as he would in a traditional laboratory because he does not understand or does not choose to perceive the content-based and pedagogical differences between the two curricula.

ULTIMATE SUCCESS!

With so many factors contributing to the curriculum as a whole, it is important to take a step back and to look at the final impact. Students experiencing the CASPiE curriculum directly express that their laboratory course exposed them to a new understanding of science. Many students even go so far as to say that they did not know what science was before CASPiE.

Many of the CASPiE students describe their experience as eye-opening regarding specifically how science is practiced. The experience of conducting their own research project brings these students to understand not only what scientists do but that they can also accomplish these tasks themselves. One student described her experience in this way:

I, um, I want to do research in the future, and I think this experience helped me to really understand that it's not, like, a magical experiment and you come up with magical data and some magical conclusion, and that it is frustrating, but you get through it, and you get over it, and you'll run it again and if it's just as frustrating, you'll do it again … it's actually made me really excited to do more research. (Student 3)

This student's realization about the true nature of science practice helped her to become more excited about research. In this way, the student came to understand not only what research was but that she could also do research herself.

Another student had a similar experience and described her understanding of the impact of CASPiE, saying,

Um, possibly in the sense that I don't think that, I didn't think that I had as much capacity for research, I suppose. I don't know if, I was kind of nervous about not being able to handle being thrown into a research type of situation. So I think that that's like, you know, broken the barrier for me, I guess, and given me capacity for research later in life. (Student 4)

This student explicitly describes how she did not think that she would be able to do research before her CASPiE experience; however, the experience with a research-based laboratory showed her that she had the skills necessary to work in a research environment.

Faculty notice the impact of CASPiE implementations at their institutions as well. One module author notes a change in students at his institution, saying,

The feedback that I've got from [the instructor] who's now sort of running the module is that the students really appreciate having the more research-based experience rather than, rather than sort of rote, just simple ah, formulaic experiments. I think they really, they get a lot more out of it. And I've certainly noticed a lot more students having, who have taken the CASPiE module um now approaching faculty members wanting to do research. That's been a, a quite significant upswing of students in our department who've actually gone on now to do research with faculty members, which is very encouraging. So, I think it gives them a taste of, a little, sort of a taste of what research is about. (Module author 4)

In his comment, this faculty member indicates the impact that CASPiE experience has had on students at his institution. He also shows his personal excitement about the change. As with Student 4's statements, he indicates that students learn what science is like and also gain self-efficacy regarding scientific research.

The complexity of integrating research elements into the laboratory environment is counterbalanced by the impact that early research experiences can have on students. Integrating research into the undergraduate curriculum is complicated and difficult at times; however, the positive impacts on students are immense. The retention of students in the sciences because of their increased research self-efficacy is reason enough to face the challenge.

CONCLUSIONS

We are striving to develop a new laboratory curriculum that provides students with a short-term authentic research experience. Our research shows that achieving this goal requires excellent instructional materials, instructor understanding of the curriculum's goals, and thorough TA preparation. When these components are present, students receive an educational experience that they can really be involved in. The curriculum gives students the autonomy to decide how best to answer their research question, and their work is relevant in a much broader context than available in traditional chemistry laboratory curricula.

The research-based approach to laboratory education achieves many educators' goals for the undergraduate laboratory curriculum (Nagda et al. 1998; Wenzel 2003; Lopatto 2004; Seymour et al. 2004), such as increasing student motivation (Module author 4) and developing an interest in the topic (Student 4). CASPiE modules place the students' laboratory work in the context of ongoing scientific research and, in turn, increase student motivation for the laboratory work. Students acquire hands-on laboratory skills as they carry out their experiments.

The CASPiE curriculum exposes students to a cross section of the scientific research experience; students in a traditional undergraduate laboratory do not have this kind of access and often fail to understand what science is like before choosing to leave the field altogether. The integrated and connected approach of teaching students the science content and the process of science simultaneously provide unique access and unparalleled benefits to the students who participate.

REFERENCES

Abraham, M. R., M. S. Cracolice, A. P. Graves, A. H. Aldhamash, J. G. Kihega, J. G. Palma Gil, and V. Varghese (1997) The nature and state of general chemistry laboratory courses offered by colleges and universities in the United States. *Journal of Chemical Education 74*(5), 591–594.

American Chemical Society (ACS) (2007) *Undergraduate Professional Education in Chemistry: Guidelines and Evaluation Procedures*, p. 23. Washington, DC: American Chemical Society.

Armstrong, H. E. (1903) *The Teaching of Scientific Method and Other Papers on Education*. New York: Macmillan and Company.

Boud, D. J., J. Dunn, T. Kennedy, and R. Rhorley (1980) The aims of science laboratory courses: A survey of students, graduates and practising scientists. *European Journal of Science Education 2*(4), 415–428.

The Center for Authentic Science Practice in Education (CASPiE) (2005) http://www.caspie.org (accessed November 30, 2009).

Chen, J., G. B. Call, E. Beyer, C. Bui, A. Cespedes, A. Chan, J. Chan, S. Chan et al. (2005) Discovery-based science education: Functional genomic dissection in *Drosophila* by undergraduate researchers. *PLoS Biology 3*(2), e59.

Hanauer, D. I., D. Jacobs-Sera, M. L. Pedulla, S. G. Cresawn, R. W. Hendrix, and G. F. Hatfull (2006) Teaching scientific inquiry. *Science 314*(5807), 1880–1881.

Hegarty-Hazel, E. (1990) The student laboratory and the science curriculum: An overview. In *The Student Laboratory and the Science Curriculum*, ed. E. Hegarty-Hazel, pp. 3–26. New York: Routledge.

Hodson, D. (1993) Re-thinking old ways: Towards a more critical approach to practical work in school science. *Studies in Science Education 22*, 85–142.

Hodson, D. (1996) Laboratory work as scientific method: Three decades of confusion and distortion. *Journal of Curriculum Studies 28*(2), 115–135.

Hofstein, A. and V. N. Lunetta (2004) The laboratory in science education: Foundations for the twenty-first century. *Science Education 8*(1), 28–54.

Lopatto, D. (2004) Survey of Undergraduate Research Experiences (SURE): First findings. *Cell Biology Education 3*(4), 270–277.

Nagda, B. A., S. R. Gregerman, J. Jonides, W. von Hippel, and J. S. Lerner (1998) Undergraduate student-faculty research partnerships affect student retention. *Review of Higher Education 22*(1), 55–72.

Russell, C. B. (2008) Development and evaluation of a research-based undergraduate laboratory curriculum. PhD dissertation, Purdue University.

Russell, C. B. and G. C. Weaver (2008) Student perceptions of the purpose and function of the laboratory in science: A grounded theory study. *International Journal for the Scholarship of Teaching and Learning 2*(2).

Seymour, E., A.-B. Hunter, S. L. Laursen, and T. Deantoni (2004) Establishing the benefits of research experiences for undergraduates in the sciences: First findings from a three-year study. *Science Education 88*(4), 493–534.

Tamir, P. (1990) Evaluation of student laboratory work and its role in developing policy. In *The Student Laboratory and the Science Curriculum*, ed. E. Hegarty-Hazel, pp. 242–266. New York: Routledge.

Trumper, R. (2003) The physics laboratory–A historical overview and future perspectives. *Science & Education 12*, 645–670.

Weaver, G. C., D. Wink, P. Varma-Nelson, F. Lytle, R. Morris, W. Fornes, C. Russell, and W. J. Boone (2006) Developing a new model to provide first and second-year undergraduates with chemistry research experience: Early findings of the Center for Authentic Science Practice in Education (CASPiE). *Chemical Educator 11*(2), 125–129.

Wenzel, T., ed. (2003) Enhancing Research in the Chemical Sciences at Predominantly Undergraduate Institutions: Recommendations of a Recent Undergraduate Research Summit. National Science Foundation. http://abacus.bates.edu/acad/depts/chemistry/twenzel/finalsummitreport.pdf (accessed November 30, 2009).

CHAPTER 11

Enriching the Chemistry Experience for All Students: Sensorial Experiments That Include Visually Challenged Students

MARIA OLIVER-HOYO

North Carolina State University, Raleigh, NC

INTRODUCTION

This book presents different faces of "inclusive learning environments" that share in common the integration of teaching approaches to the learning of a diverse student population. My own description adds several criteria when considering a learning environment an "inclusive" one such as the following:

1. It provides the same opportunities to all students, including full participation, so that the experiences may be equally rewarding.
2. The ultimate goal is to make instruction effective, meaning that it results in a stimulating, engaging, and productive learning experience.
3. Achievement is measured at two equally important levels: the academic or scholar level and the social or practical level.

This chapter highlights how the "sensorial" experiments we have developed for chemistry comply with these criteria. Sensorial experiments utilize multiple human senses, particularly the senses of smell,

Any opinions, findings, and conclusions or recommendations expressed in this material are those of the author(s) and do not necessarily reflect the views of the National Science Foundation.

hearing, touch, and sight when studying chemical processes and when performing chemical techniques. By relying on more than one sense, it becomes possible to make observations or determinations even when experiencing difficulties with one sense in particular. Chemistry laboratory experiences predominantly utilize the sense of sight while neglecting other human senses. These alternative experiments appeal to students with sensorial difficulties such as visually impaired students since these students are fully capable to use their senses of smell, touch, or hearing. Therefore, the sensorial experiments provide an opportunity to integrate visually impaired students into the laboratory experience in an active and independent way. By adding multiple senses to the arsenal of available resources, these experiments also add richness to the chemistry experience of all students, regardless of physical abilities.

The focus of our efforts is the inclusion of visually impaired students in the chemistry laboratory experience. Even though chemistry is the physical science that studies everything we perceive through our senses, only the sense of sight is exploited in chemical investigations, leaving students with visual impairments at a particular disadvantage in chemistry laboratories. Statistics from the American Foundation for the Blind show a U.S. population of ~10 million blind and visually impaired people of whom ~93,600 are students and 55,200 are legally blind children (American Foundation for the Blind). Schooling of this population is markedly different with 45% of people with these conditions earning a high school diploma versus 80% of fully sighted individuals earning this degree. The 1998 National Science Foundation (NSF) report on Women, Minorities, and Persons with Disabilities (PWD) in Science and Engineering clearly stated that "students with disabilities are significantly underrepresented in undergraduate and graduate majors in science, mathematics and engineering" (NSF 1998). Seymour and Hunter (1998) suggest three reasons responsible for these statistics, which are faculty attitudes, financial aid requirements, and allotted time for completing the degree. I must add one more reason to this list that pertains to chemistry, and it is the lack of instructional materials, specifically chemistry experiments, that instructors may easily get and implement. As an experimental science, the laboratory component is essential in chemistry courses. Depriving students of such an experience is like teaching one how to swim on dry land.

A LOOK AT AVAILABLE LABORATORY RESOURCES

The American Chemical Society (ACS) Chemists with Disabilities (CWD) Committee have two publications that address disability issues in general (Miner et al. 2001; ACS). *Working with Disabilities* is an inspirational account of 17 chemists' lives and professional endeavors. The 2001 edition of *Teaching Chemistry to Students with Disabilities: A Manual for High Schools, Colleges, and Graduate Programs* presents information related to all aspects of instruction including mentoring and advocacy and the universal design for accessibility. Both publications are available in print and online. In general, the resources available are plenty and are very informative. Since there is a wide range of disabilities, the search for useful information can be overwhelming to the instructor who just received news that a student with a disability is enrolled in their course. It is important to narrow the search toward the specific disability of your student and to embrace the journey as potentially the most enlightening learning experience for you as a teacher. The resources provided in this section pertain primarily to visual challenges specific to chemistry laboratory instruction.

Much of what has been done to address visual impairments in science laboratories may be classified into one or more of three categories: structural modifications, tactile learning aids, and specialized equipment or instrumentation (to include assistive technologies). Structural modifications are aimed to provide easy access to working areas and to comply with safety standards. Examples in this category include ramps, oversized doorways, and tables, fume hoods, and eye washes with adjusted heights. This category is fully supported by federal legislation such as the Architectural Barriers Act (ABA), Americans with Disabilities Act (ADA), and Individuals with Disabilities Education Act (IDEA; U.S. Department of Education, Office of Special Education Programs), among others. The only requirement needed to familiarize visually impaired students with the facilities and laboratory organization is some extra time and prior planning so that students may visit the laboratory beforehand, may be able to identify specific needs, and may orient themselves in the new environment.

Tactile learning aids generally involve converting a drawing, figure, or graph into a raised form of the graphics. Low-cost tactile drawings can be obtained using a tracing wheel on a sheet of Braille paper, or a hot glue gun to emphasize lines or shapes (Supalo 2005). Wikki Stix, a pliable waxed yarn, was creatively used by Poon and Ovadia (2008) to produce raised markings for organic instruction. More sophisticated technologies available include "swell paper," "heat pens," and "thermoform diagrams" that work with heat-reactive materials (Supalo 2005; National Center for Tactile Diagrams). A thermoform diagram results when a sheet of plastic is heated and vacuumed on top of a template. Braille graphics produce images consisting of dots that represent specific combinations of letters, words, punctuation, or numbers. A traditional molecular model kit was effectively modified as a tactile learning aid to predict reaction products, synthetic routes, mechanisms, and nomenclature in an organic chemistry course (Poon and Ovadia 2008). Laboratory materials that incorporate these tactile aids promote self-sufficiency into the learning process of visually impaired students.

Sophisticated learning aid models have been produced using stereolithography. Images obtained from a scanning probe microscope (SPM) were used to produce 3-D physical models via "layering manufacturing" (Jones). This technique also produced tactile phase diagrams. Laser stereolithography utilizes a light-sensitive liquid polymer that hardens as a laser cures the material in the shape of an image (Venanzi et al. 1996). The accuracy of this method in representing positions, angles, and diameters of atoms surpasses those obtained from traditional molecular model kits. A promising goal for Skawinski et al. is to incorporate electric charge distributions in these 3-D molecular models. However, these are examples where cost and ready availability are crucial issues to consider, and the success of these types of efforts should be measured by the extent of their accessibility.

There are promising technologies that should be explored for laboratory chemistry instruction such as "Smelly Vision," "Tactile Colour," and "Tactile-Audio" (National Center for Tactile Diagrams). Smelly Vision uses colored paper sheets that have been impregnated with different aromas, while Tactile Colour uses texture to denote different colors. Tactile-Audio requires a computer to produce audio information as different regions of the graphic are pressed. These technologies could be exploited to attract the attention of the student toward specific portions of relevant information or to describe chemical processes and their sequential changes.

Adaptations to equipment or instrumentation may be categorized into low versus high technology. Simple modifications to existing laboratory equipment include raised markings on measuring devices, spoons with sliding covers, bands of sandpaper as warning symbols for labeling hazardous materials, and notched dial settings (Tombaugh 1981; Supalo 2005, 2008). Automatic dispensers for volume measurements include auto pipettes, automatic fluid dispensers, and syringes. Talking instruments produce an audible output and available ones include thermometers, scales, stopwatches, calculators, and multimeters such as pH meters and voltmeters. The talking beaker failed to be useful as its precision, amounts measured, and preset liquid densities were not in line with typical chemistry tasks (Neely 2007). Ratliff (1997) measured conductivity of elements and solutions using sound rather than the traditional LED setup for an inexpensive and easy-to-assemble audible conductivity testing apparatus. Light probes have been used for decades to assist the visually impaired student in determining liquid levels such as in the meniscus of a solution, in the color changes of solutions, or in the formation of precipitates (Tombaugh 1981). Light probes emit a tone that increases in pitch proportional to changes in light intensity. A common light sensor consists of a small photocell in a case with an opening in one end in order to allow light in. A continuous tone is emitted and the pitch of the sound increases with the brightness of the light hitting the opening. A newer version of a light probe is a submersible audible light sensor tested with color changes of typical titrations (Supalo 2008). We have used a commercial light probe to "hear" changes in color, solution viscosity, and the formation of precipitates in an organic qualitative analysis laboratory (Bromfield-Lee and Oliver-Hoyo 2007). The LumiTest light probe is manufactured by CareTec and may be obtained from the American Printing House for the Blind (catalog #1-03956-00).

Computer assistive technologies for students with visual impairments include not only Braille note takers and output programs but also a wide range of sophisticated programs that allow the visually impaired student to access information in a variety of formats. Examples of these include optical character recognition (OCR), speech recognition, and voice output programs as well as screen readers, screen enlargements, and print text enlargement functions. Neely (2007) reported the pros and cons of some technological modifications to assist visually impaired students in the laboratory, which include a magnification camera device and digital readouts on common laboratory equipment. A variety of keyboard aids are available for easier access to computer keys. The University of Washington Disabilities, Opportunities, Internetworking, and Technology (DO-IT) Program has compiled a summary of this kind of available resources for the visually impaired (Access Technology Lab at the University of Washington). "Smart" instruments that can be interfaced to computers include electronic balances and UV–vis spectrophotometers and many laboratory programs come with a variety of sensors such as colorimeters, pH electrodes, pressure sensors, and conductivity probes. These assistive technologies provide access to computing resources that could be crucial in laboratory instruction. The availability of these assistive technologies vary greatly from one institution to another, and it is the responsibility of instructors and visually impaired students to find out what is in fact available and what resources could be made available for their instruction.

The sense of touch used in the "nanoManipulator" provides students with the opportunity to explore the microscopic world as it uses a microcomputer programmed to control a joystick device, which gives tangible response from an atomic force microscope (AFM) image (Jones et al. 2003). Even though its initial application has been for studying viruses, the learning gains shown by students in this study indicate that this tool is beneficial for studying the morphology of microscopic dimensions; therefore, its potential as a useful component in chemistry laboratories is very promising. MolySim is a molecular modeling software that connects a physical model with a virtual display of its properties including the dynamic energetics of the system. Haptic technology has also been used in advanced visualization systems such as the Molecular Visualiser (MV), a cost-effective tool to study molecular quantum dynamics (Davies et al. 2005). Further development of instructional materials for chemistry using these novel approaches would offer immense learning benefits to all students and an opportunity to level the field for visually impaired students studying chemistry.

Students experiencing sensorial challenges and their instructors' input, interest, and experience are the driving forces behind the development of tactile learning aids and adaptations of equipment. It should be pointed out that looking at reactions from a different perspective is all it might take to incorporate a visually impaired student into the laboratory experience. As an example, there are no adaptations to equipment or instrumentation or assistive technologies involved in an activity where reactions that generate carbon dioxide are observed by blind students via sound and touch (Gettys et al. 2000).

"Tactile observation of all apparatus is the one general pedagogical rule for blind science instruction" (Cetera 1983). However, an easy mistake to make is to think that what is felt using the sense of touch equates what is seen with the eyes (Mayo 2004). Mayo's research showed that perspectives of visually impaired students are different from those of sighted individuals. When embossed figures of molecular geometries drawn as if seen from the side were shown to blind participants, their mental images of the representation were described and drawn as if they had looked at it from the top. The students often confused what the figure represented as a side of the molecule for the top of the molecule. Therefore, translating mental images, evoked from reading a text, into a two-dimensional drawing was problematic for the participants. As a result, the blind participants had difficulties drawing, describing their drawing perspective, and choosing symbols to transmit their mental images into a two-dimensional drawing.

SENSORIAL EXPERIMENTS

The use of multiple senses in chemistry experiments represents a more effective approach at mentally stimulating students and provides a better opportunity for memory retention as suggested by the effects of multisensory input in virtual environment experiments (Dinh et al. 1999). Research shows that olfactory stimuli enhance memory recognition (Cann and Ross 1989) and that once an aroma is retained, there is long-term resistance to distortion (Engen and Ross 1973). Coupling the sense of smell to hearing and touch provides a powerful combination to explore chemical principles in the laboratory.

Pioneers

Exemplar sensorial experiments have served as the foundation of and inspiration for our work (Rancke-Madsen and Krogh 1956; Hiemenz and Pfeiffer 1972; Tallamn 1978; Lunney and Morrison 1981; Flair and Setzer 1990; Wood and Eddy 1996). The sense of hearing has been utilized in potentiometric (Tallamn 1978; Lunney and Morrison 1981) and conductometric titrations (Hiemenz and Pfeiffer 1972). A voltage-controlled oscillator provided a tone whose pitch changed as changes in the pH occurred. In a potentiometric titration, a large potential change per increment of titrant indicates that the end point is approaching. Scanning the audible titration, the student found the end point since a large potential change produced a louder tone. A special piston burette was constructed and attached to a retroreflective sensor allowing graduation marks to be detected as the thumbwheel delivered the titrant. A computer attached to the system counted the graduation marks that passed through the detector and calculated the volume delivered under the command from the student. In a conductometric titration, the variation in conductance was used to detect the end point of a titration since the end point of a conductometric titration is found by extrapolating the two linear portions of a conductance versus volume of titrant added (Hiemenz and Pfeiffer 1972). When the discontinuity at the equivalence point was sharp enough, the audio indicator was able to judge the null point of the bridge. The output of the electrode attached to the conductivity bridge was passed through an amplifier to a loudspeaker instead of activating the normal eye-type null detector. The titration consisted of adding a titrant until a minimum in sound intensity was perceived. These authors also reported an easier way of identifying the end point of a titration using a "blank" titration. In such, the amplifier gain control was adjusted to a barely audible signal when electrodes were immersed in a solution titrated to the phenolphthalein end point. Electrodes were then placed immediately in an unknown solution and were titrated until the minimum signal was perceived. Aiming toward a specific sound made this titration comparable to the color change of phenolphthalein. This seminal work showed that to the visually impaired, variations in sound perceptions are reliable indicators of the progress in a chemical reaction. However, when special instrumentation is involved, there is the risk of limited accessibility. In addition, when high technology adaptations are part of the setup, these require constant updates adding to the cost of making them available to the few students and their instructors who might need them. For example, David Lunney's (1994) piston-driven burette model is no longer manufactured.

In the previous examples, the progress of the experiments was monitored via hearing rather than sight, and the chemistry involved ran its course unchanged. Wood and Eddy (1996) took a very different approach as they manipulated the chemistry of aroma release in onions to utilize the vegetable as an olfactory indicator. The raw onion aroma can be "arrested" in basic solutions, and it is then detected under neutral or acidic conditions. When titrating a basic solution that is odorless and to which drops of onion extract have been added, the onion odor first becomes noticeable when neutrality is reached. Flair and Setzer (1990) evaluated other potential olfactory indicators and found eugenol, obtained from cloves, to be viable. The characteristic fragrance disappeared when eugenol was rebasified, making back titrations feasible. As far back as the 1950s, Rancke-Madsen and Krogh (1956) reported accurate olfactory indicators such as sodium butyrate, sodium sulfide, quinoline, pyridine, and ammonium chloride for strong acid/base titrations. In all these experiments, the onset of an aroma determined the end point of the titrations.

The titrimetric modifications described above exemplify either an instrumental adaptation (as in the case of the potentiometric and conductometric titrations) or a chemical manipulation (olfactory indicators). The first uses the sense of hearing, while the latter type appeals to the sense of smell. Unfortunately, instrumental adaptations that utilize one-of-a-kind homemade instruments are not readily available to the typical educator or may even be out of circulation. Also considered is the fact that at institutions where the enrollment of handicapped people is low, the justification of specialized equipment can be very difficult to obtain. On the other hand, the chemical manipulations discussed rely on readily available "chemicals" (onions and cloves) and represent comparable costs and laboratory preparation times as traditional titrimetric experiments. Since all students may perform these types of titrations, full participation side by side of sighted and visually impaired students is possible.

Our Contributions

The ultimate goal for the development of these sensorial experiments is for them to be useful to students with visual impairments and their instructors. All resources already discussed in this chapter may be incorporated to provide students with these challenges an opportunity to perform chemistry as independently as possible. In addition, the benefits and role of a sighted assistance are well documented (Cetera 1983; Miner et al. 2001), and these assistants too could get a different perspective about the science that studies everything we perceive through our senses.

The sensorial experiments presented in this chapter have been thoroughly tested in undergraduate laboratories for safety, accuracy, and reproducibility. Since students use senses not commonly involved in the laboratory setting, techniques such as wafting the chemicals instead of direct inhaling of the aromas must be demonstrated and continuously monitored. The chemicals used were thoroughly screened for toxicity levels, inhalation thresholds, aroma offensiveness, and general hazards. In addition, these experiments are conducted at a small scale for easier handling and to reduce the impact of potential accidents. Accurate scientific results are crucial for these experiments to be pedagogically valid. Results of these experiments are comparable to traditional experiments allowing students to draw the same conclusions as topic-related traditional experiments. The priority is for students to investigate chemical principles using multiple senses rather than using equipment or machines to conduct a procedure. These sensorial experiments are designed as inquiry based rather than as verification laboratories, and students need data from other classmates in order to reach their own conclusions. These practices promote active thinking and collaborative work.

Some practical issues worth pointing out include availability and expense of chemicals, aroma control, and aroma fatigue. In many cases, such as when using onions and garlic as olfactory indicators, "chemicals" come in the form of readily available consumer products. For aroma control, fume hood use is maximized whether it is needed for safety reasons or not, and all scented waste is disposed of outside the laboratory. Aroma fatigue is avoided via different strategies including "purging" the sense of smell using coffee crystals between titration runs or the use of a mini fan to disperse transient outbursts of aromas.

The sequence of sensorial experiments discussed below represents the evolution of our approaches at developing a series of experiments to fit into the first two years of undergraduate chemistry courses, general and organic.

Olfactory Titrations   Our first attempt at developing sensorial experiments focused on controlling the chemistry of a reaction in order to utilize the onset of an aroma as the indicator for the end point of a titration (Neppel et al. 2005). Previous results on olfactory indicators were intriguing from a chemical standpoint (Flair and Setzer 1990; Wood and Eddy 1996). The mechanism for aroma release in onions is very similar to garlic since these belong to the same plant genus, *Allium*. The literature pointed us in the right direction indicating the conditions that favor the production of aroma precursors, which included low pressure, exposure to light, temperatures above 50°C, and an acidic medium. It was the pH dependence that we manipulated in order to successfully use garlic as an olfactory indicator. The raw garlic odor or onion aroma is perceived under neutral or acidic conditions but is "arrested" and not perceived under basic conditions. Even though onions have been utilized as olfactory indicators (Wood and Eddy 1996), the end-point results were precise but were not accurate, deviating consistently from the actual standardized concentrations. On the other hand, even though garlic is perceived more often as having an unpleasant aroma, it gave very accurate results and therefore is the plant of choice for this type of olfactory titrations. The procedure is simple and straightforward. The plants are chopped (onions) or crushed (in the case of garlic) in a sodium hydroxide (NaOH) solution and are allowed to soak for 10–15 minutes to quench their characteristic aromas. A pipetted portion of the plant solution is used as the analyte and is titrated with hydrochloric acid, HCl to the olfactory end point.

A very different source, vanillin, has also given accurate results. A saturated solution of vanillin is prepared by dissolving vanillin in ethanol and by adding a small portion of this solution to the basic analyte solution, NaOH. This type of olfactory titration is more appealing to students due to the pleasant aroma of the vanillin. Instructors might find this titration more attractive as well since the concentration of the NaOH is only 1 M versus the 2 M required to quench the aromas of onions and garlic. A detailed account of the chemistry involved, procedures, and student handouts is available (Neppel et al. 2005).

Qualitative Organic Analysis   Qualitative organic schemes are as common in organic laboratory courses as titrations are in general chemistry laboratories and the study of the chemistry of functional groups is fundamental. So our next approach was to take a commonly performed organic qualitative laboratory and to transform it into a multisensorial one. Students categorize a series of compounds by their aromas and use this classification to determine which confirmatory tests are required to identify their unknown (Bromfield-Lee and Oliver-Hoyo 2007). Students work in groups and together decide on the confirmatory tests they should perform on their unknowns, bringing the collaborative aspect to central focus. The confirmatory tests are common; however, the way to interpret and observe the results are not. Each confirmatory test result involved at least two senses and one, testing for alcohols, involves all four senses allowed in laboratory instruction. For example, the generation of heat relies in the sense of touch, while the generation of a gas relies on hearing the gas bubbling formation. Interesting and unexpected outcomes include the differences in the perception of aromas by gender. For example, 28% of the males could not detect a methanol aroma, while no females had such difficulty; and 25% of males did not detect ethanamide, while only 8% of the females expressed such difficulty. These findings point out the relevance of using multiple senses in the laboratory as a resource for truly "inclusive learning environments."

Kinetics Building upon and connecting chemistry topics via a laboratory experiment are ideal opportunities for strengthening pedagogical aims. Kinetics is a topic we start discussing in general chemistry and revisit in almost every core chemistry course. Since our targeted academic time frame is the first two years of undergraduate education, two sensorial experiments were developed that fit into the general or the organic curriculum.

Acetyl Salicylic Acid (ASA) Hydrolysis Kinetics Traditional experiments involving the kinetics of the compound ASA, the main ingredient of commercial aspirin tablets, usually depend on UV spectrometry to monitor the concentration of ASA in the UV region of the spectrum or the visible absorbance of the colored complex formed between iron or copper with ASA (Street 1988; Borer and Barry 2000). An alternative method to studying ASA kinetics spectroscopically is by monitoring the production of vinegar from the acidic hydrolysis of ASA via the onset of the vinegar smell. Vinegar's aroma is well recognized by students and they note the time taken for the emergence of the vinegar aroma for various concentrations of ASA. These data are then used to find the order of the reaction as students plot the data into characteristic kinetic graphs and find that plotting log rate (log 1/t) versus log [ASA] indicates a first-order reaction. For this experiment, we chose the acidic hydrolysis of ASA due to its direct production of vinegar and the convenient speed of the reaction.

The procedure is uncomplicated as students measure certain amounts of ASA (1–5 g) into beakers that are placed in water baths kept at ~65°C. A 1 M solution of HCl (20 mL) is added to the beaker and the start time is recorded. The final time is that when the vinegar aroma is observed and persists for 30 seconds. The time elapsed for the emergence of the aroma to appear is used for plotting log rate versus log [ASA]. The most crucial aspect of the procedure is the incorporation of a mini fan in the setup of the experiment (Fig. 11.1). This fan must be positioned so that the airflow passes through the top of the beaker (avoiding fast cooling of the flask itself) and is at a constant distance from the reaction beaker (~12 in.) for safe detection of the aroma and consistency. This is a very similar setup as the one we utilized with olfactory titrations (Neppel et al. 2005). This procedure entails simple measurements of a solid amount and a solution volume, monitoring of the water bath temperature, and recording of time elapsed for the emergence of the vinegar aroma. With the available equipment modifications such as volume dispensers, talking scales, and thermometers,

FIGURE 11.1 Setup for an ASA kinetic experiment.

the visually impaired student would easily perform the entire procedure independently.

Esterification Kinetics   Sulfuric acid-catalyzed Fischer esterifications are a standard reaction studied in lecture and laboratory organic courses, and for which a combinatorial approach to produce a variety of esters has been used for instructional purposes (Birney and Starnes 1999). We exploited the fact that many esters have naturally occurring, pleasant and easily recognized aromas to study the effects of molecular structure, concentration of reagents, and amount of catalyst in the condensation reaction of certain carboxylic acids with alcohols (Bromfield-Lee and Oliver-Hoyo, in press). In the first part of the experiment, students synthesize different esters individually and characterize the aromas obtained. In the second part, students study in groups the structural effects on the kinetics of the reaction so that they could later elucidate the effects of chain lengths and branching. Each group is assigned a carboxylic acid to react with four different alcohols. Finally, students study how the concentration of acid and the amounts of alcohol and catalyst affects the rate at which the characteristic esters' aromas emerge. Data collection involves individual work, group work, and class collaboration.

Thermometric Titrations   Touch is a sensitive human sense as we may perceive slight variations in texture, temperature, and consistency as examples. However, it is not a sense that by itself we could use to make quantitative determinations in a laboratory. Thermometers help us make those measurements and talking thermometers permit visually impaired students to monitor the temperature of a reaction precisely and quantitatively. We have monitored acid/base titrations by the heat changes produced and determined the end point of these thermometric titrations from the highest temperature achieved. The experiment may be designed for different groups of students titrating different acids with the same strong base, such as $1\,M$ NaOH, as all these reactions are exothermic. We have successfully used hydrochloric, sulfuric, nitric, and acetic acids obtaining comparable results as with a colorimetric phenolphthalein indicator. As the common practice in titrations dictates, the first titration is done to identify the volume range where the most dramatic temperature drop occurs. In two additional trials, students slowly add a titrant over this range and collect precise data of the temperature and volume of titrant added. The same successful thermometric results were obtained with the redox titration of potassium permanganate (oxidizing agent) with iron II, Fe^{2+} (reducing agent,

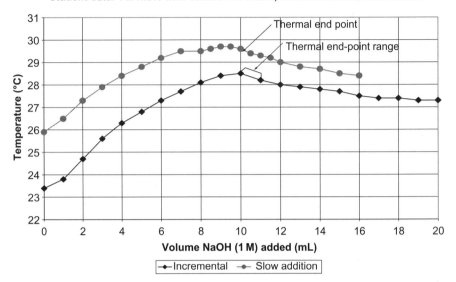

Student data: 1 M nitric acid versus 1 M NaOH, incremental and slow addition

FIGURE 11.2 Rough and precise end points of the thermometric titration of HNO$_3$ with NaOH.

Fe(NH$_4$)$_2$(SO$_4$)$_2$·6H$_2$O). Interestingly, this redox titration produces a color change without an extrinsic colorimetric indicator. In order to prevent sighted students to use the color change as a reference, the titration vessel may be covered, leaving small holes to place the thermometer and the burette. Graphs of temperature versus volume of base easily show accompanying thermal changes, and tracking the curves with, for example, a hot glue gun makes them available for visually impaired students to study by themselves. A typical thermometric graph is shown in Fig. 11.2.

CONCLUDING REMARKS

Bogner et al. (2006) have pointed out that in order to "see with the *mind's eye*" requires stimuli that elicit familiar things. Nothing becomes "familiar" until sufficient exposure has taken place. These sensorial experiments provide the means to expose visually impaired students to chemistry experimentation and to become familiar with chemical techniques and procedures. The academic success of students with disabilities is higher when this population is included in traditional classrooms and is provided the opportunity to engage in active learning (Baker et al. 1994; Watson and Johnston 2004; Kumar et al. 2001). The importance and benefits of all students doing exactly the same thing has been documented (Baker et al. 1994; McDaniel et al. 1994; Kumar et al. 2001). This chapter has discussed alternative and enhancing methods for chemistry laboratory instruction.

Attitudinal barriers are considered by many to be the main problem in the search for better opportunities for people with disabilities. These barriers include, among others, lower expectations that result from an outdated understanding of disability and fear or inappropriate concerns about safety. Society seems to equate a disability with reduced abilities overall, or to focus on the disability rather than on the intrinsic abilities of the individual. Augmenting the learning experience via sensorial channels is a viable way to enhance the teaching and learning processes for all students. Changing to more inclusive laboratory activities (those that can be used by all students) has the potential of decreasing the attention to differences between students with and without disabilities and of reducing the attitudinal barriers faced by students with visual challenges.

Lawrence Scadden (2000), a successful blind scientist, gives a candid account on the impact assistive technologies have brought to the visually impaired and emphasizes the fact that every technology originally developed specifically to assist people with disabilities has eventually benefited everyone. For example, a 1934 invention known as the "readphone," which reproduced literature and music on long-playing disks, was first used by The American Foundation for the Blind and by the Library of Congress to develop talking books. Later, this invention led to the recording of long-playing records for the general public. Scadden summarizes his "formula for success" as "tools, training, and opportunities." Tools are constantly being developed as emerging assistive technologies and sensorial experiments exemplify. Training depends primarily on the motivation and interest sought after by affected parties. However, for opportunities to become available, the academic enterprise should work together at the individual (students and instructors) and collective (departments, administration, and funding agencies) levels if "inclusive learning environments" would be available for students with visual impairments. It is the collective efforts that give training and opportunities a chance to exist.

Need unleashes *creativity* and *resourcefulness*. An appropriate and excellent example came to my attention when one of my colleagues forwarded an e-mail from Harry, a general chemistry instructor who had a totally color-blind student who could only distinguish shades of gray but with great accuracy. After some conversations, the student offered to build a table of standard colors to use as reference. And the story goes ...

> It turns out that his father was a confetti salesman (among other things) and so the student brought me a packet of sample confetti sheets that included every known color. At the beginning of the lab, I would identify which pieces of confetti looked like the critical colors in the lab, and he would then work until he had the same shade of gray in the flask, etc. as was on the sample sheet. To my surprise, his analytical results were excellent! I still remember watching him doing titrations, holding the pack of confetti up against the flask, titrating until the two gray colors matched. If I learned nothing else from that experience, I came away convinced that students can teach me a lot if I will only listen to them.

As this chapter has highlighted, sensorial experiment procedures allow students with visual impairments to actively participate in chemistry laboratories, along with nondisabled classmates. The instructional materials developed require minimal need of peripheral equipment, have comparable preparation times and cost to traditional laboratories, and are properly designed to fulfill safety requirements. The incorporation of available technologies into these experiments makes it possible to adapt these chemistry experiments to the needs of students with sensorial disabilities. As these experiments are disseminated and implemented in chemistry laboratories, the opportunity to encourage the interest of students with visual impairments to study chemistry increases. The prospect is to change mistaken views about the potential, inherent in all, to contribute and enjoy science.

REFERENCES

Access Technology Lab at the University of Washington. http://www.washington.edu/computing/atl/DOCS/atl2.html#visual (accessed December 14, 2009).

American Chemical Society (ACS). Working chemists with disabilities: Expanding opportunities in science. http://membership.acs.org/C/CWD/workchem/start.htm (accessed December 14, 2009).

American Foundation for the Blind. Statistical snapshots. http://www.afb.org/Section.asp?SectionID=15#num (accessed December 14, 2009).

Americans with Disabilities Act (ADA). http://www.ada.gov/ (accessed December 14, 2009).

Baker, E., M. Wang, and H. Walberg (1994) The effects of inclusion on learning. *Educational Leadership 52*, 33–35.

Birney, D. M. and S. D. Starnes (1999) Parallel combinatorial esterification. *Journal of Chemical Education 76*(11), 1560–1561.

Bogner, D., B. L. Wentworth, and J. Ristvey (2006) Our place in the spongy universe. *Science Teacher 73*(3), 38–43.

Borer, L. L. and E. Barry (2000) Experiments with aspirin. *Journal of Chemical Education 77*(3), 354–355.

Bromfield-Lee, D. C. and M. T. Oliver-Hoyo (2007) A qualitative organic analysis that exploits the senses of smell, touch, and sound. *Journal of Chemical Education 84*(12), 1976–1978.

Bromfield-Lee, D. and M. T. Oliver-Hoyo (2009) An esterification kinetics experiment that relies on the sense of smell. *Journal of Chemical Education 86*(1), 82–84.

Cann, A. and D. Ross (1989) Olfactory stimuli as context cues in human memory. American Journal of Psychology *102*(1), 91–102.

Cetera, M. M. (1983) Laboratory adaptations for visually impaired students: Thirty years in review. *Journal of College Science Teaching 12*(6), 384–393.

Davies, R. A., N. W. John, J. N. MacDonald, and K. H. Hughes (2005) 3D technologies for the world wide web. In *Proceedings of the Tenth International Conference on 3D Web Technology*, pp. 143–150.

Dinh, H., N. Walker, C. S. Kobayashi, and L. A. Hodges (1999) Evaluating the importance of multi-sensory input on memory and the sense of presence in virtual environments. In *Proceedings of the IEEE Virtual Reality Conference*, http://www.informatik.uni-trier.de/~ley/db/conf/vr/vr1999.html (accessed December 14, 2009).

Engen, T. and B. M. Ross (1973) Long-term memory of odors with and without verbal descriptions. *Journal of Experimental Psychology 100*(2), 221–227.

Flair, M. and W. N. Setzer (1990) An olfactory indicator for acid-base titrations. *Journal of Chemical Education 67*(9), 795–796.

Gettys, N.S., E. K. Jacobson, and J. Bell (2000) More than meets the eye: Nonvisual observations in chemistry. *Journal of Chemical Education 77*(9), 1104A–1104B.

Hiemenz, P. and E. Pfeiffer (1972) A general chemistry experiment for the blind. *Journal of Chemical Education 49*, 263–265.

Jones, G. M., T. Andre, R. Superfine, and R. Taylor (2003) Learning at the nanoscale: The impact of students' use of remote microscopy on concepts of viruses, scale, and microscopy. *Journal of Research in Science Teaching 40*(3), 303–322.

Jones, R. R. Scientific visualization through tactile feedback for visually impaired students. http://www.dinf.ne.jp/doc/english/Us_Eu/conf/csun_98/csun98_170.html (accessed December 14, 2009).

Kumar, D. D., R. Ramasamy, and G. P. Stefanich (2001) Science for students with visual impairment: Teaching suggestions and policy implications for secondary educators. *Electronic Journal of Science Education 5*(3). http://

ejse.southwestern.edu/original%20site/manuscripts/v5n3/articles/art03_kumar/kumar.html (accessed December 14, 2009).

Lunney, D. (1994) Data acquisition for the visually impaired student. *Journal of Chemical Education 71*, 308.

Lunney, D. and R. Morrison (1981) High technology lab aids for visually handicapped chemistry students. *Journal of Chemical Education 58*, 228–231.

Mayo, P. M. (2004) Assessment on the impact chemistry text and figures have on visually impaired students' learning. Doctoral dissertation. Purdue University, West Lafayette, IN.

McDaniel, N., G. Wolfe, C. Mahaffy, and J. Toggins (1994) The modification of general chemistry labs for use by students with disabilities. *Journal of Rehabilitation 60*(3), 26–29.

Miner, D., R. Nieman, A. B. Swanson, and M. Woods (2001) *Teaching Chemistry to Students with Disabilities: A Manual for High Schools, Colleges, and Graduate Programs*, 4th ed. Online by ACS Committee on Chemists with Disabilities. http://membership.acs.org/C/CWD/TeachChem4.pdf (accessed December 14, 2009).

MolySym. MolySym: Technology for research and education. http://www.molysym.com/. Product info at http://www.nsti.org/Nanotech2004/showabstract.html?absno=873 (accessed December 14, 2009).

National Center for Tactile Diagrams. Making tactile graphics. http://www.nctd.org.uk/MakingTG/index.asp, http://www.nctd.org.uk/MakingTG/craft.asp (accessed December 14, 2009).

National Science Foundation (NSF) (1998) Women, minorities, and persons with disabilities in science and engineering. Report at http://www.nsf.gov/statistics/nsf99338/; quote at http://www.nsf.gov/statistics/nsf99338/pdf/c3.pdf (accessed December 14, 2009).

Neely, M. B. (2007) Using technology and other assistive strategies to aid students with disabilities in performing chemistry lab tasks. *Journal of Chemical Education 84*(10), 1697–1701.

Neppel, K., M. T. Oliver-Hoyo, C. Queen, and N. Reed (2005) A closer look at olfactory titrations. *Journal of Chemical Education 82*(4), 607–610.

Poon, T. and R. Ovadia (2008) Using tactile learning aids for students with visual impairments in a first-semester organic chemistry course. *Journal of Chemical Education 85*(2), 240–242.

Rancke-Madsen, E. and J. A. Krogh (1956) The use of the senses of taste and smell in determining the end point of an acid-base titration. *Acta Chemica Scandinavica 10*, 495–499.

Ratliff, J. L. (1997) Chemistry for the visually impaired. *Journal of Chemical Education 74*(6), 710–711.

Scadden, L. (2000) Soaring expectations. College of the Pacific Convocation. http://www.colorado.edu/sacs/ATconference/scadden1.htm (accessed December 14, 2009).

Seymour, E. and A. Hunter (1998) *Talking about Disabilities: The Education and Work Experiences for Graduates and Undergraduates with Disabilities in Science, Mathematics, and Engineering Majors.* Washington, DC: American Association for the Advancement of Science.

Skawinski, W. J., T. J. Busanic, A. D. Ofsievich, T. J. Venanzi, V. B. Luzhkov, and C. A. Venanzi (1995) The application of stereolithography to the fabrication of accurate molecular models. *Journal of Molecular Graphics 13*(2), 126–134.

Street, K. W. (1988) Method development for analysis of aspirin tablets. *Journal of Chemical Education 65*(10), 914–915.

Supalo, C. (2005) Techniques to enhance instructors' teaching effectiveness with chemistry students who are blind or visually impaired. *Journal of Chemical Education 82*(10), 1513–1518.

Supalo, C. (2008) Low-cost laboratory adaptations for precollege students who are blind or visually impaired. *Journal of Chemical Education 85*(2), 243–247.

Tallamn, D. E. (1978) A pH titration apparatus for the blind student. *Journal of Chemical Education 55*(9), 605–606.

Tombaugh, D. (1981) Chemistry and the visually impaired. *Journal of Chemical Education 58*(3), 222–225.

U.S. Department of Education, Office of Special Education Programs. IDEA Website. http://idea.ed.gov (accessed December 14, 2009).

Venanzi, C. A., W. J. Skawinski, and A. D. Ofsievich (1996) Molecular models by laser stereolithography. In *Physical Supramolecular Chemistry*, eds. L. Echegoyen and A. E. Kaifer, pp. 127–142. Dordrecht: Kluwer Academic Publishers. Second version: The Use of Laser Stereolithography to Produce Three-Dimensional Tactile Molecular Models for blind and Visually Impaired Scientists and Students. http://people.rit.edu/easi/itd/itdv01n4/article6.htm (accessed December 14, 2009).

Watson, S. W. and L. Johnston (2004) Teaching science to the visually impaired. *Science Teacher 71*(6), 30–35.

Wood, J. T. and R. Eddy (1996) Olfactory titration. *Journal of Chemical Education 73*(3), 257–258.

CHAPTER 12

Media in Chemistry Education

WILLIAM J. DONOVAN

University of Akron, Akron, OH

This chapter discusses media including interactive websites and response systems, commonly called "clickers." Both types of media have become ubiquitous in the chemistry classroom in recent years, and both have shown great promise in increasing student engagement and success. Students with diverse backgrounds and learning styles can use these media to increase their understanding of chemistry concepts alone and in collaboration with other students, tutors, and teaching assistants (TAs), and their instructor.

LITERATURE ON LEARNING WITH MEDIA

Interactive websites and clickers are forms of instructional media. In his review of the literature, Kozma (1991) described a framework in which the learner actively collaborates with the medium to construct knowledge, where the learner participates in an active process of construction based on management of cognitive resources and extraction of information from the environment. Kozma's framework, which he described as "the learner actively collaborating with the medium to construct knowledge," is compatible with both distributed cognition and constructivism (Vygotsky 1978; Bodner 1986; Fosnot 1996; Howe 1996). Active collaboration with the medium is consistent with the

Making Chemistry Relevant: Strategies for Including All Students in A Learner-Sensitive Classroom Environment, Edited by Sharmistha Basu-Dutt
Copyright © 2010 John Wiley & Sons, Inc.

distribution of functioning over more than just the person but rather over the person and the medium. Collaboration or distribution means that what the learners experience and how they will learn will depend on the characteristics of the medium.

Kozma's review was a response to Clark (1983), who asserted that "media do not influence learning under any conditions ... Media are mere vehicles that deliver instruction but do not influence student achievement any more than the truck that delivers our groceries causes changes in our nutrition." According to Clark, the mode of presentation may affect the cost or extent of the distribution of education, but "only the content of the vehicle can influence achievement" (p. 445). Clark cited many reviews and studies that concluded that media do not affect learning and that many studies that compared media had been fruitless.

Salomon and Clark (1977) called many studies of media "research *with* media," meaning that some other issue was really at the heart of any differences observed between groups of students. According to Clark, some studies used the medium as a conveyance of some other treatment, and sometimes the difference in medium was mistakenly given credit for any benefits to students. "Media attributes, rather than media per se, were dealt with, and a strong need for conceptualizing these attributes in terms of their effects and functions was suggested," wrote Salomon and Clark (1977, p. 116).

Clark (1987) and Salomon (1979) continued to recommend that researchers study attributes of media and how these attributes affect the processing of information in learning. Clark described the promise of an approach based on media attributes based on three expectations:

- that the attributes are an integral part of media and would provide connections between learning and the uses of media;
- that attributes of media would "provide for the cultivation of cognitive skills for learners who needed them"; and
- that identified attributes would provide unique independent variables with causal connections from media to learning.

According to Clark, only the second expectation was fulfilled, implying that studies of media attributes may contribute to instructional design but not to theory development.

According to Salomon (1979), there are several classes of attributes that cut across media: the contents conveyed by the medium, the symbol systems used to convey the contents, the technologies used, and the situations in which the medium is used (p. 14). Salomon states that "although each medium consists of all four classes of attributes, not all attributes matter equally in learning. In some cases, a technological attribute distinguishes one medium from others in terms of the learning experiences it affords" (p. 14). This is congruent with a distributed cognition model. Salomon's interest in the attributes of a medium and what it affords continued; Salomon (1993a) went on to edit a volume on distributed cognition.

Clark argued that media do not affect learning; he argued that it is the content rather than the mode of presentation that makes the difference. In many students' experiences, the best way to learn a concept can depend on the medium. Some people have difficulty understanding certain concepts when presented in one way who immediately grasp the concept when presented in another way. Computers and the Web can be used to present materials in ways that books cannot. Clickers and the pedagogy accompanying them allow students to think and rethink their answers to questions in class and to discuss their conceptions with other students. The theory of affordances deals with what a medium is "good for" or how it can best be used, which is what Salomon described in his discussion of technological attributes. A website has different affordances than a book or a set of lecture notes, and the learner may use the affordances that he or she finds useful to learn. Kozma's model of the "individual collaborating with the medium" is very important.

Distributed Cognition

In the distributed cognition model, knowledge is not only in the mind but is also rather spread out or distributed among the components of a system. The system may include the learner as well as tools the learner uses, whether these tools are symbol systems or artifacts. In this model, learning is the change in patterns of activity or interactions between the learner and the tools. Perkins (1993) called the combination of the person and some external components the "person-plus," taking as an example unit of analysis a student and the student's notebook. "We could say that this person-plus system has learned something, and part of what the system has learned resides in the notebook rather than in the mind of the student," Perkins wrote. In the person-plus paradigm, the system "learns" or "knows" something rather than the knowledge residing solely in the person.

This model could be applied to a situation in which a student uses web-based materials to learn a chemistry topic. The student and the medium can together engage in learning the material, especially if the student finds the medium to be useful in representing the chemistry in ways that the student cannot picture well or reproduce well in traditional ways.

Pea (1993) discussed the use of tools as they relate to distributed cognition. He differentiated between a primarily quantitative change in functioning and a more qualitative change in mental functioning made by tools. According to Pea, computer tools serve not as "amplifiers of cognition but as reorganizers of mental functioning." Amplification would refer to quantitative changes in functioning or accomplishment (possibly more of the same type of functioning), whereas reorganization implies that the actual structure of the function is changed. Pea describes the use of computer tools to change the way that people learn. This means that it is possible for the final outcome to be the same using different tools, but the mental functioning that takes place in the process can be completely different, depending on the selection of the tool used.

In a discussion of several papers on distributed cognition in *Educational Psychology Review*, Moore and Rocklin (1998) call for greater precision in the use of the term "distributed cognition." They point out that there is a wide range of interpretations of the term, varying from "conceptualization of cognition as an individual phenomenon that is influenced by factors external to the individual, to a conceptualization of cognition as a social phenomenon that cannot be reduced to individual psychological constructs." Some authors cited by Moore and Rocklin treat distributed cognition as a purely social type of cognition and do not leave much room for the individual aspect of cognition. Also, Moore and Rocklin state their concern that many authors believe that the more cognition is distributed, the better the system will function. Distributed cognition models cognition as distributed over people and artifacts rather than characterizing cognition as only in the head. It does not necessarily follow, argue Moore and Rocklin, that more distribution leads to better cognitive system function. Citing the medical profession as an instance of more distribution not being better in the long run, the authors make the analogy of a doctor, medical artifacts, and his/her patient in a community. Clearly, the medical knowledge necessary to make a diagnosis is best not distributed too widely across the system.

Salomon (1993a,b) argues that there may be some cases where cognitions are not distributed, and some classes of cognitions, such as higher-order knowledge, that cannot be distributed. These cognitions can only remain in the individual. Another concern that Salomon states is that focusing only on the distributed cognition does not allow for any change or growth in the cognitions. The role of the individual must be considered, as distributed cognitions and individual cognitions must both be present in order to be able to affect and to develop each other.

Salomon also stated that in education, the design of situations of distributed cognitions should be designed to promote or scaffold the "cultivation of individuals' competencies." This is an important issue to consider in the design of web-based materials that offer students new or different ways to learn chemistry. It is valuable to offer students the opportunity to use materials that may help them learn the chemistry, and properly designed materials will offer students the ability to learn while being supported by the special affordances of the Web.

Affordance Perceptions

Gibson developed the theory of affordances in studies of visual perception, differentiating between the physical world and the environment. According to Gibson (1977), the physical world is considered to be the "reality" of what surrounds an organism. The environment is taken to be the surroundings as perceived by the organism.

The theory of affordances is related to the theory of distributed cognition as well as to the various types of media discussed by Kozma (1991). Gibson defined affordance as "a specific combination of the properties of its substance and its surfaces taken with reference to an animal" and stated that "the affordances of the environment are what it offers the animal, what it provides or furnishes, either for good or evil" (Gibson 1979). A definition that is somewhat more useful in the context of media in education was presented by Pea (1993): "'Affordance' refers to the perceived and actual properties of a thing … that determine just how the thing could possibly be used."

Gibson states that while an affordance consists of physical properties in reference to an animal, the affordances do not depend on the animal. The affordances of an object do not depend on the observer. But Gibson contends that affordances of an object are not objective if objectivity means that the object has no reference to any animal. According to Gibson, the affordances of an object are facts of the environment but not at the level of physics, leaving out animals. This argument is parallel to the argument that a tree falling in the forest when no one is there to hear it still makes a sound. An object has its own affordances even if the observer is not there to perceive them. A complication to this thought is that in affordances involving humans with free will versus animals, the affordances that are perceived may vary from person to person.

The Web offers affordances to students that more traditional means of teaching cannot offer, such as the use of three-dimensional user-manipulable representations of molecular structures and animated molecular-level representations. While the use of on-screen three-dimensional representations may be specific to computers running web browsers with the appropriate browser plug-in, the website offers the affordances that a textbook, handout, or page of lecture notes cannot offer. The Web offers the user the ability to use types of representations that require affordances not offered by paper. The affordances of the computer are well matched to the subject matter involved, which is visual in nature and requires the use of a medium offering the appropriate affordances.

In addition, the Web can offer users animated representations of chemical phenomena and animations at the molecular and particulate levels. Animated representations can show molecular motion and connections among macroscopic, microscopic, and symbolic worlds. On a website, molecular-level views of motion can be shown as appropriate for the respective phases of matter. On paper, it is not possible to show such motion in progress.

Visualization

Experts typically are able to work in several representational domains, while novices are more limited in flexibility between or among representational domains. Kozma et al. (1993) raise a critical point with regard to this use of representations: that without the information that experts know, novices may construct internal representations that rely on and are constrained by the external qualities of the representation. "For example, it may be that because the color stops changing in the observed representation, students develop the misconception that at equilibrium, reactions come to a stop" (1993, p. 7).

Kozma et al. state that each type of external representation provides some unique information and shares some information with other representations. Paivio's (1990) dual coding theory may explain why the linkage of multiple types of representations is useful and helps the learner store information in memory more efficiently. Dual coding theory states that texts are processed and encoded in verbal systems, but pictures and graphics are processed and encoded in image and visual systems in the learner. Paivio's experiments showed that the imagery code and the verbal code could be used to store and retrieve information. Finke (1989) gives examples of different types of words that require the use of the different codings. Concrete objects such as "table" or "horse" are coded primarily with imagery, whereas abstract concepts such as "truth" or "beauty" are coded with verbal codes. The presence of a separate imagery code allows for recall of information that was never explicitly coded verbally. Also, Paivio's dual coding theory explains why pictures are frequently easier to remember than words. One can use verbal and imagery codes to recall a picture, while one can only use verbal codes to recall a word.

According to Paivio (1990), dual coding theory has a hierarchical conceptual structure. At the most general level are symbolic systems, cognitive systems that serve as a symbolic or representational function. At the next level, we divide verbal and nonverbal symbolic systems, and at the next level, these systems expand into sensorimotor (visual and auditory) subsystems. The lowest level of the hierarchy consists of the representational units of the system, which Paivio referred to as *logogens* and *imagens*. Logogens are associated with the verbal system, while imagens are associated with the nonverbal system. Words activate logogens, while objects or their pictures activate imagens. Interactive websites that include visualization and various types of representations for chemical phenomena and structures allow students to take advantage of the dual coding, using both verbal and imagery codes to recall and to understand chemical concepts and phenomena.

Using clickers in classes allows students to discuss their ideas with other students during discussion time, and the questions posed while using clickers can take advantage of the same websites. Students with different learning styles (Felder 1993) are served well by the give-and-take with students with other learning styles. Visualization of the macroscopic, microscopic, and symbolic worlds simultaneously also can help students understand the relationships between these worlds, which chemists take for granted but which many students find challenging. Technology such as clickers can also allow feedback to the instructor similar to that provided by one-minute papers, allow students to identify parts of diagrams, or respond to questions over previous readings (Felder 1993). These opportunities to collaborate with other students and with the medium are consistent with the distributed cognition model, where the knowledge is spread among the components of the system. Of special interest in Felder's paper is the point that addressing the needs of diverse learners can help students in both the tier who will go on to STEM major fields and the tier who initially intend to but change fields, described by Tobias (1990). It is doubtless that taking steps to address differences in learning styles and to increase feedback among students and to the instructor will lead to greater inclusion and success for all.

INTERACTIVE WEBSITES

Interactive websites have been shown to be beneficial to females and to weaker students. Nakhleh et al. (2000) evaluated the Educational

Materials for Organic Chemistry (EMOC) website at Michigan State University. Students in several undergraduate organic chemistry courses used the materials on the EMOC site in their course work. Students' attitudes toward website use were generally positive, and students generally believed that using the website helped them learn the chemistry. The researchers gathered from the results of the focus group interviews that weaker students appear to be more attracted to EMOC than stronger students. The researchers speculated that differences in motivation may be the cause for the weaker students being more attracted to the website than stronger students. The website nonuser interviewed exhibited good mastery of the chemistry in her interview, while the website nonusers interviewed showed some weaknesses in their understanding.

In a related study (Donovan and Nakhleh 2001, 2007), statistically significant differences between genders were observed for student responses to several of the scaled-response survey questions regarding website use. In each case where statistically significant differences existed between the mean female and male responses, the mean female response was more positive toward website use. This trend was observed in more agreement with positive statements (such as "I liked using the web page because it allowed me to work at my own pace") as well as disagreement with negative statements (such as "I would rather NOT use the Web to learn chemistry"). In a previous study of the use of computers in biology laboratories (Eichinger et al. 2000), a similar result was found: in each case where gender differences existed in student responses, female students responded more favorably about computer use than male students. This congruence of opinion presents interesting possibilities with respect to how such materials might be used to improve science courses for women. The availability of computers and web-based tutorial materials might be a positive aspect of a science course for women.

Donovan and Nakhleh reported that many students used the interactive tutorial website because of the visualization that was possible using it, because of access to materials and information, and because of needing help with chemistry. The special affordances that the Web offers, such as the ability to display molecules in three dimensions and to rotate the representations of the molecules to different positions, allow the visualization that is not available in a book or on a chalkboard or piece of paper. Using these representations was also useful to students as a reinforcement tool and as an aid in studying.

The most common reason given for not using the website was that the student had forgotten about the website or did not know about it in the first place. Students who knew about the website but did not use it often said that they understood the chemistry material and did not need to use the website to help their understanding.

Analysis of student interviews showed that those who used the website generally made more incomplete statements and fewer correct statements in the discussion of coordination chemistry than students who did not use the website. In the overall comparison of the students' concept maps, the proportions of correct, incorrect, and incomplete statements were almost the same. The students who used the website made overall twice as many links on their concept maps as website nonusers made. This may mean that students who used the website may be thinking about chemistry more or differently and are thus making more statements.

The primary reason students gave for not using the website was that they did not know about the website or that they forgot about it. This is unfortunate but is a very real problem. Among students who did know about the website, the majority of students who chose not to use it stated that they did not need to use the website to learn chemistry. Therefore, it seems a reasonable recommendation that instructors assign the website as part of the students' homework or other assignments, so all students have the opportunity to use the Web and to see for themselves if it would be useful to them.

Some students found that using the website benefited them. In particular, website use appears to be useful for students who wish to visualize the chemistry involved. According to the students, the multimedia and visual aspects of the website present a useful way to gain understanding of chemistry. Therefore, it may be useful to students for the instructor to use or to demonstrate these web-based representations in class when covering relevant material. The author recommends that these findings be taken into consideration in the design and implementation of web-based materials for chemistry courses. The availability of the multimedia materials on the website allows the instructor to show three-dimensional representations and animations in lecture and reminds the students that the materials are available for review or for further study at another time. The ability to use web-based tutorial materials in class seems to be a valuable opportunity to expose students to a medium that may help them learn chemistry.

It is worth noting that the first generation of websites demonstrating three-dimensional molecular structures commonly used the Chime plug-in, which must be installed on the student's computer or on the campus computers, while the newest generation uses Jmol animations and structures that live on the web server itself and can be viewed by anyone who has Java installed on his or her computer. This may require an additional level of sophistication in the development of the websites, since the Chime molecules can be generated by software such as Hyperchem, but in the author's experience, there are sufficient undergraduate students in computer science and even in chemistry with enough computer experience to be of great help in developing such materials for the Web, for credit or pay.

Textbook publishers are now moving toward making portable the media content on the websites that accompany texts. For example, recent editions of some textbooks (Burdge 2009; Silberberg 2009) are marketed featuring companion content online that can be downloaded as an MPEG file to a student's media player (e.g., an iPod) or to the hard drive of a computer. Thus, a student does not need to actually be online in order to view this type of media; the student can download it and view it later on their portable media player or computer.

Media on Disk

Not all visualization tools have to be delivered by Internet. As described in the next chapter of this book, Gabriela Weaver's group at Purdue University has developed a multimedia DVD for physical chemistry (Dyer et al. 2007; Jennings et al. 2007). The DVD format allows students to have all of the material on the disk without having to download it, and Weaver's DVD focuses on research using sophisticated instrumentation that students at many institutions may not have access to and would otherwise not encounter at all. Users of the DVD gained understanding of the research areas that they otherwise would not have encountered, demonstrating real-world applications and practical uses of the physical chemistry content in their course. The evaluation of the physical chemistry DVD showed that students using the DVD materials showed significant learning gains, increased recognition of applications of physical chemistry to real-world problems, and increased interest in further study (Jennings et al. 2007).

CLICKERS

The use of electronic response systems (clickers) in the chemistry classroom has increased greatly in the past several years. Possibly made famous in the "Ask the Audience" lifeline on *Who Wants to Be a Millionaire?*, clickers have been used in many classroom settings at all grade levels. While increasingly common in chemistry, response systems are also commonly used in other science courses such as geology (McConnell et al. 2006), astronomy (Duncan 2006), computer science (Cutts et al. 2004), physics (Burnstein and Lederman 2001, 2003), medicine (O'Connor et al. 2006), and biology (Preszler et al. 2007), as well as other disciplines such as economics (Hinde and Hunt 2006), law (Burton 2006; Kift 2006), and mathematics (Pelton and Pelton 2006). The use of these systems in college chemistry classes has been the subject of symposia at meetings including the Spring 2005 American Chemical Society National Meeting and the 2006 and 2008 Biennial Conferences on Chemical Education. DeHaan (2005) cites clickers as a key technology-based innovation for the classroom, and Educause (2005) states that "clickers—plus well-designed questions—provide an easy-to-implement mechanism. Clicker technology enables more effective, more efficient, and more engaging education."

Response systems are commonly used with conceptests, which are part of the peer instruction pedagogy described by Mazur (1997). The use of conceptests with a clicker system in chemistry has been investigated with respect to its relationship to web-based quiz assignments (Bunce et al. 2006), and grades have been compared between semesters using and those not using clickers (Hall et al. 2005). Asirvatham (2008) has used clickers in improvement of understanding of molecular visualization. A review of literature (Judson and Sawada 2002) pointed to the instructor's pedagogical practices, not the response system technology itself, as the key to student comprehension. Another very complete review of recent literature on applications and research on clicker use in chemistry classes (MacArthur and Jones 2008) summarized 56 papers and stated that student response to clicker use is positive, while the effect on learning is positive in some cases and inconclusive in others.

Peer instruction increases student interaction and de-emphasizes the instructor-led lecture by allowing students to discuss concepts and to help each other learn. Conceptests (Landis et al. 2001) are short questions covering the material just discussed by the instructor. Students may select an answer to the conceptest by a show of hands, by colored or numbered cards, or with a response system. If the overall response to the conceptest shows that many members of the class do not understand the concept, the instructor may ask the students to discuss the question with each other. In the discussion, students are encouraged to try to convince their neighbor of their own answer but might also change their own minds if they see shortcomings in their own original answers. The class revotes, and hopefully the answer distribution has improved with respect to the number of correct answers. Chemistry conceptest questions are available online (University of Wisconsin

2006) and in print (Landis et al. 2001), and many textbook publishers supply banks of questions with the ancillary instructors' materials.

It is important to note that the pedagogy of conceptests is not necessarily linked to the use of an electronic response system, but that the response system allows the instructor to collect and save the data and allows students to retain a degree of anonymity when they respond. Judson and Sawada's (2002) review of literature in the *Journal of Computers in Mathematics and Science Teaching* on response systems in college lecture halls noted that professional development for instructors using or considering the use of a response system should focus on pedagogy.

Donovan (2008) reported that student understanding improved between in-class clicker questions and paired exam questions on many topics, verifying student reports that their understanding was aided by the conceptests and clicker use in class. In addition, King and Joshi (2008) reported that a higher percentage of female students than male students actively participated in lectures using clickers and that active students (with activity defined through participation using clickers) in lecture of both genders performed better on exams than the nonactive students. This shows that, as for website use, it is possible that females may benefit from the interactive technology of clickers in their chemistry class.

Types of Clicker Systems: Hardware and Software

Clicker systems are offered by numerous vendors, and the hardware that students use can be purchased by the students themselves (and brought to class every day) or by the institution (and checked out to students in class and returned at the end of class). The latter was more commonplace in the early days of clicker systems, with some early response systems actually consisting of keypads built into lecture hall seating armrests, but the most recent clicker systems use pads that are owned by individual students and are brought to class every day. The clicker may often be used in more than one course, with the student taking the clicker to each course.

The earliest wireless clicker systems involved infrared (IR) signals from the clickers to the receiver, while more modern clicker systems use radio frequency (RF) signals. Disadvantages of the IR systems include the limited number of signals that can be received at a time by the receiver and the line-of-sight issues specific to IR signals; the clicker must be aimed directly at the receiver in order for the signal to be sent and received. RF clickers have fewer such issues, with much faster data collection and line-of-sight issues eliminated. With the increased capability of RF clickers, however, usually comes a higher price to the student. The latest generations of many clicker systems also offer input of numeric and symbolic answers, allowing many more answer choices.

The hardware that the instructor uses in class can vary as well. The earliest IR clicker system used at the author's institution involved an IR receiver that plugged into a computer via a serial or USB port; the USB connection required a special connector cable since the serial connection was the native configuration. Large classes had difficulty with the single clicker receiver, so a combining box was installed, which combined the signals sent to two receivers installed at opposite ends of the chalkboard and projection screen at the front of the lecture hall. Students were directed to aim their clicker toward the receiver on their side of the room. As the wired receivers were difficult to maintain and the wires would occasionally be disconnected for unknown reasons, a wireless receiver system was later installed in the room, with wireless receivers in the positions previously occupied by the wired models, and an additional wireless receiver in the back of the lecture hall. Thus, students toward the back of the room were asked to aim their clicker toward that receiver when responding. While the motion of aiming the clicker backward looked awkward, there were by far fewer signal collisions, and thus the data were collected much more quickly. Barber and Njus (2007) cite similar growing pains with IR clickers in large lecture halls.

The wired receiver system was retired when the RF system was adopted for use campus-wide. The RF system currently used on the author's campus uses a small receiver unit (about the size of a box of chalk) that plugs in to the USB port via a wire. A new version of the RF receiver, which resembles a USB flash drive in size and appearance, is now available. The RF receiver does not have to be set up so as to allow the students to aim their clickers at it, but it does seem that the best results are obtained when the receiver is pointed toward the students, with no obstructions such as books blocking the receiver.

At the author's institution, the initial clicker usage began at the departmental level, with several different systems in use and students being issued clickers in class, which they returned every day. The university's Institute for Teaching and Learning later ran a pilot program that paid for the activation of IR clickers for students in courses participating in an internal grant program. Other courses could also use the same clicker system, but the clicker activation would have to be paid for by the students. In both cases, students purchased the IR clickers for approximately $4 in the bookstore. After the introduction of

RF systems, the university signed a contract with one clicker provider for several years, after taking bids from several providers. The contract allows students to use their clicker in any number of courses during each semester, and for any number of students to use the clicker system. The provider has supplied receiver hardware and software for instructors to use. The students purchase the RF clickers, which cost approximately $20 at the bookstore but can be used for multiple courses and can theoretically be used for a student's whole four years (or more) at the university.

The bookstores sell the clickers for several reasons. First, it is more convenient for the students to purchase a clicker there then at the campus computer store since the bookstore is set up at the book rush time at the beginning of the semester for the large numbers of students who need clickers for their various classes. Second, students who have scholarship funds that can pay for books and class materials can have their scholarship funds applied through the campus bookstore, while the computer store cannot process such purchases. Third, the bookstore can more easily deal with any issues such as replacing defective clickers since they are staffed by a larger group and can more easily process returns and exchanges without inconveniencing other customers. Also, the bookstores on the author's campus deal in used clickers just as they do books, buying used clickers from students at one fraction of their value and selling the used clickers to students at a slightly higher fraction of the new value. When the campus changed from the first generation of RF clickers to the second, the university purchased a number of the first-generation clickers to make up a pool of loaner clickers to be used for nonclass activities including orientation, seminars, teacher workshops, training sessions, and meetings.

The software that operates the clicker system has improved in functionality over the years, with added features and support for PowerPoint. As the author does not generally use PowerPoint (but rather writes on partially filled notes on a tablet computer), he does not use all of the features, but the generally most essential features of recording student answers to questions and generating class attendance reports have been present from the earlier versions of the software onward. Setting up a course in the software used by the author involves using an alphanumeric class key unique to the university online, and then having the students register their own clickers into the system while entering the serial number of their clicker. The entry of the serial number allows the student's clicker to be identified by the receiver and to be connected to the student's record in the software.

In the event that a student mistypes the serial number, their clicker is not recognized. In that case, the student must return to the website to re-register the clicker with the correct serial number. It is possible for the instructor to modify the clicker serial number in his own computer, but his experience has been that when the roster is automatically downloaded and updated at the next class period, the old, mistyped serial number from the online system will overwrite the revised one. This is done by design; if students lose their clicker and buy a replacement, they can go to the website and delete their old serial number and enter their new one, and the new serial number information will be updated into the instructor's roster at the next class period if the instructor has enabled the automatic roster update function. Of course, this also means that mistyped serial numbers must be corrected online by students.

Self-registration of clickers by students can be done in a structured setting if students have a class meeting in a room with a computer for each student. Some universities have students meet in a computer laboratory room for recitation sections, while others have computers for data collection in the laboratories. In these cases, students can be asked to bring their clicker to class the first day the recitation or laboratory meets, and the teaching assistant (TA) can walk the students through the registration process. Interestingly, Griff and Matter (2008) found that for a biology course where students self-registered their clickers on their own, course grade was correlated with the order of clicker registration. Students who registered their clickers early had a higher probability of success in the course than those registering their clickers later. It would be an interesting study to see if a similar relationship exists for chemistry. In the case of the author of this chapter, it might be possible to arrange the students in order of clicker registration, since the first student is assigned number 1, and so on, and to see if there are more high grades in the low clicker numbers.

Some students invariably forget to bring their clickers to class. The author does not adjust the attendance record manually in the clicker software for such situations. While there is an attendance and participation component to the grade in the course, the student can miss up to 20% of the class periods without penalty. Students who are not present and are not participating for at least 50% of the class periods, however, receive no credit for the attendance/participation portion of the grade. The allowance for missing some classes is intended to account for "life happening"—flat tires, oversleeping, and so on—as well as forgotten clickers. The author will tell students who forgot their clicker and are very concerned to send him an e-mail so he knows about the situation and can check at the end of the semester to see if it makes a difference in the student's grade, but it rarely does. Occasionally, a student will have a cell phone out when telling the author about their forgotten clicker, and the author usually points out to the student that they did bring their cell phone to class (even though they did not need it in class), but they did not bring the clicker (which they do need). One such observation is usually enough to get such students to remember to bring their clicker in the future.

Another issue in the logistics of clicker use is battery life. Students will frequently report by the end of the second semester that their clicker's batteries have died, especially if the clicker was used in several classes. The author carries a supply of AA batteries (purchased during a buy one-get two free sale) for this situation; students are usually very pleasantly surprised to find out that they do not have to pay their instructor back for the batteries. The author has found that it simplifies life to have the students be able to replace their batteries and to participate immediately rather than worry about how to handle the attendance/participation grading for students who were present with a clicker with dead batteries. Since the batteries may die unexpectedly, it is not fair to penalize these students, and providing new batteries takes care of that issue. It is worth noting that the clickers used in the author's class turn off automatically after going out of range of the receiver unit after class, or can be turned off manually. The particular clickers do not turn off automatically after a certain period of time in class when the receiver is present, however, which is a good thing—otherwise, the clickers might turn off if there is a long lag between questions.

Barber and Njus (2007) mention a point that is worth noting: Macintosh support for clicker software generally lags behind that for PC users. The author of this chapter uses a PC, but his colleagues who teach organic chemistry are Mac users, and one of those colleagues used clickers in his organic chemistry class for a semester but did not continue due primarily to logistical difficulties in the classroom (he does not use PowerPoint, and the layout of the room permitted use of either the board or the projection screen at one time, but not both) but also because the Macintosh software did not function as well as that for the PC at the time, and the hardware was more difficult to install and use. Since that time, the Mac support has improved greatly, but experience of users on the author's campus and of others at other universities using other systems suggests that Mac support for clickers still lags behind PC support.

Asking Questions and Getting Answers

The details of how one enters questions ahead of time in clicker software or in presentation software such as PowerPoint are varied, depending on the specific provider and indeed on the specific version of the software, and thus will not be discussed here. However, in nearly all cases, it is possible to enter questions ahead of time, whether in the PowerPoint presentation or in the clicker software itself. Whether one method is better is a matter of personal taste. It is possible that the clicker software may have better support than PowerPoint for the various symbols, arrows, superscripts and subscripts, and diagrams that are required for many common chemistry questions, but it is also possible that the clicker software may be more limited in this regard. The author uses a tablet computer and writes on Journal files produced from Word documents, and thus writes the questions on the document, often on a blank page to leave room for working the solution after the question is done.

Many clicker software programs allow the instructor to not show the histogram of responses to a conceptest question by default. This is useful for situations where the instructor would like the students to vote and then to discuss their answer with a neighbor for a minute, then to revote, following the peer instruction model. However, it is more common to show the histogram from the first vote to see if discussion is necessary.

If a large number of students (more than 75%) have the correct answer, further movement toward the correct answer may be simply due to the other students moving toward what they see as the most popular answer. In this case, it may be sufficient for the instructor to discuss the problem and its solution briefly, or to allow a student who answered correctly to do so, before moving on. Even if 95% of the students in the class have answered correctly, it is important to discuss the problem and its solution for the benefit of the 5% who answered incorrectly, as well as those who may have answered correctly only by a guess and who do not really understand the concept. The instructor should always make sure that the students have a grasp on the concept before moving on.

If a small number of students (less than 30%) have answered correctly, it may not be fruitful for discussion to occur, as it is not likely that many students will discuss the question with a classmate who answered the question correctly. It is common in the author's class to ask a question that the instructor knows will receive such a response pattern, to get the students' attention. For example, after a discussion of acid–base equilibria, including the cases of very dilute solutions in which the simplifying assumption of excluding water dissociation does not work, the author asks the class what the pH of a 1.0×10^{-8} M solution of nitric acid would be, to the nearest pH unit. The most common responses are usually 8 and 6, with 8 being more popular. Very few students answer with the correct response of pH 7. When the answer distribution appears and is discussed, the students are very surprised to find out that over 90% of the class was wrong! This is quite an attention-getter, and the students' interest in why the answer was 7 is piqued. The author usually finds and asks a student who answered 6 why 8 would not be a possible answer (as the solution contains an acid), and the response from the majority who answered 8 is usually a sigh and some comments to the effect of "d'oh, that's right!"

Some users of clickers have developed methods for going beyond the typical multiple choice and numeric entry of answers by using the input features in innovative ways. Sauers and Morrison (2007) described the use of clickers with numeric entry capability to enter the reagents, in order, that would need to be used to accomplish a particular transformation in an organic synthesis. Ruder (2008) described a similar use of numeric input to describe the movement of electrons in arrow-pushing mechanisms.

The author, inspired by these ideas, has used the numeric entry in similar ways. For example, for questions where there are two whole numbers required for an answer, the whole numbers can be entered separated by a decimal point. Questions such as "how many sigma and pi bonds are present in this molecule?" or "how many alpha and beta decays occur in the decay of uranium-238 to radon-222?" can be answered by students by entering the first number, then a decimal point, then the second number. For example, the correct answer to the question regarding uranium-238 decay would be entered as "4.2." Numeric input can also be used to allow selection of multiple items. The author frequently asks a question regarding the molecular polarity of three isomers of $C_2H_2Cl_2$, drawn and numbered 1, 2, and 3. The question asks which molecule(s) would have a net nonzero dipole moment. Students are told to enter the numbers of the molecules they select. If a student chooses molecules 1 and 3, they would enter "13." If a student chooses only molecule 2, they would simply enter "2." When asking such a question, the instructor tells the students that if they believe that none of the choices are correct, they should enter "0." This strategy eliminates nearly endless lists of multiple-choice options (a), (b), (c), (a) and (b), (a) and (c), and so on, and takes advantage of the function of the clicker.

Woelk (2008) described two major classes of clicker questions, each with subclasses. The "I Am" category includes questions in the subcategories "I Am Here" (attendance generation), "I Am Prepared" (questions over reading from the day before), and "I Am Interested" (questions with more than one possible answer or educated guess questions to pique students' interest). The "I Do" category includes three subcategories: "I Learn" includes the majority of clicker questions, primarily those over material just discussed. "I Understand" questions are more conceptual and aim to separate students who truly understand the concept from those who have not yet gained a grasp on the concept. "I Apply" questions involve applications and may possibly require more information than is given in the problem statement. Woelk describes one more category, "I Will," which includes open-ended questions followed up at the beginning of the next class session. A combination of these question types will get students interested and engaged in the material.

Beyond Clickers

In the future, it is likely that clickers will be relics of an age gone by. Students already have cell phones and other devices, which may assume the place of clickers. As their costs decline, PDAs, tablet computers, and other technology may take the place of the clickers that students use today. It is possible, too, that the next innovation may be as yet unknown. Not long ago, it was not thought that the features of modern clicker systems would be available in every classroom on a campus.

SUMMARY

Interactive websites and clickers offer students ways to interact with the material and with other students in ways that traditional media do not offer. These interactions allow students with diverse learning styles the chance to see different representations of chemical phenomena and structures, and to interact with students with different learning styles than their own in class. Research suggests that interactive websites and clickers can be beneficial to women in science classes. Clickers offer feedback to the instructor and to the students during class, so that instructors may spend more time on topics that students are having difficulty with and less on those concepts that are understood, and so that students may see where they need to work on improving their understanding. Since students have different learning styles and process information differently, it is very useful for students to have the opportunity to use interactive websites on their own and in collaboration with classmates and tutors, and to have in-class discussions with other students as well as possibly with peer tutors, TAs, or the instructor. These opportunities allow the students to find other representations and understandings of the concepts that may help them along to a deeper or more complete understanding themselves.

REFERENCES

Asirvatham, M. R. (2008) Enhancing pedagogical benefits of clickers in large classrooms using information from general chemistry concept surveys. *Abstracts of Papers of the American Chemical Society 235*, CHED-1560.

Barber, M. and D. Njus (2007) Clicker evolution: Seeking intelligent design. *CBE Life Sciences Education 6*, 1–8.

Bodner, G. M. (1986) Constructivism: A theory of knowledge. *Journal of Chemical Education 63*(10), 873–878.

Bunce, D. M., J. R. VandenPlas, and K. L. Havanki (2006) Comparing the effectiveness on student achievement of a student response system versus online WebCT quizzes. *Journal of Chemical Education 83*(3), 488–493.

Burdge, J. (2009) *Chemistry*. New York: McGraw-Hill.

Burnstein, R. A. and L. M. Lederman (2001) Using wireless keypads in lecture classes. *The Physics Teacher 39*, 8–11.

Burnstein, R. A. and L. M. Lederman (2003) Comparison of different commercial keypad systems. *The Physics Teacher 41*, 272–275.

Burton, K. (2006) The trial of an audience response system to facilitate problem-based learning in legal education. In *Audience Response Systems in Higher Education: Applications and Cases*, ed. D. A. Banks, pp. 265–275. Hershey, PA: Information Science Publishing.

Clark, R. E. (1983) Reconsidering research on learning from media. *Review of Educational Research 53*(4), 445–459.

Clark, R. E. (1987) Which technology for what purpose? The state of the argument about research on learning from media. Paper presented at the Annual Conference of the Association for Educational Communications and Technology, Atlanta, GA, February 21–March 1, 1987.

Cutts, Q., A. Carbone, and K. van Haaster (2004) Using an electronic voting system to promote active reflection on coursework feedback. Proceedings of the International Conference on Computers in Education, Melbourne, Australia, November 30–December 3, 2004.

DeHaan, R. L. (2005) The impending revolution in undergraduate science education. *Journal of Science Education and Technology 14*(2), 253–269.

Donovan, W. J. (2008) An electronic response system and conceptests in general chemistry courses. *Journal of Computers in Mathematics and Science Teaching* 27(4), 369–389.

Donovan, W. J. and M. B. Nakhleh (2001) Students' use of web-based tutorial materials and their understanding of chemistry concepts. *Journal of Chemical Education* 78(7), 975–980.

Donovan, W. J. and M. B. Nakhleh (2007) Student use of web-based tutorial materials and their understanding of chemistry concepts. *Journal of Computers in Mathematics and Science Teaching* 26(4), 291–327.

Duncan, D. (2006) Clickers: A new teaching aid with exceptional promise. *Astronomy Education Review* 5(1), 70–88.

Dyer, J. U., M. Towns, and G. C. Weaver (2007) Physical chemistry in practice: Evaluation of DVD modules. *Journal of Science Education and Technology* 16, 431–442.

Educause (2005) 7 things you should know about clickers. http://connect. educause.edu/Library/ELI/7ThingsYouShouldKnowAbout/39379?time=1231652450 (accessed April 1, 2009).

Eichinger, D. C., M. B. Nakhleh, and D. L. Auberry (2000) Evaluating computer lab modules for large biology courses. *Journal of Computers in Mathematics and Science Teaching* 19(3), 253–276.

Felder, R. (1993) Reaching the second tier: Learning and teaching styles in college science education. *Journal of College Science Teaching* 23(5), 286–290.

Finke, R. A. (1989) *Principles of Mental Imagery*. Cambridge, MA: MIT Press.

Fosnot, C. T. (1996) *Constructivism: A Psychological Theory of Learning*. New York: Teachers College Press.

Gibson, E. J. (1977) The theory of affordances. In *Perceiving, Acting, and Knowing*, eds. R. B. Shaw and J. Bransford. Hillsdale, NJ: Erlbaum.

Gibson, E. J. (1979) *The Ecological Approach to Visual Perception*. Boston: Houghton-Mifflin.

Griff, E. R. and S. F. Matter (2008) Early identification of at-risk students using a personal response system. *British Journal of Educational Technology* 39(6), 1124–1130.

Hall, R. H., M. L. Thomas, H. L. Collier, and M. G. Hilgers (2005) A student response system for increasing engagement, motivation, and learning in high enrollment lectures. Proceedings of the Eleventh Americas Conference on Information Systems, Omaha, NE, August 11–14, 2005.

Hinde, K. and A. Hunt (2006) Using the personal response system to enhance student learning: Some evidence from teaching economics. In *Audience Response Systems in Higher Education: Applications and Cases*, ed. D. A. Banks, pp. 140–154. Hershey, PA: Information Science Publishing.

Howe, A. C. (1996) Development of science concepts within a Vygotskian framework. *Science Education* 80(1), 35–51.

Jennings, K. T., E. K. Epp, and G. C. Weaver (2007) Use of a multimedia DVD for Physical Chemistry: Analysis of its effectiveness for teaching content and applications to current research and its impact on student views of physical chemistry. *Chemistry Education Research and Practice* 8(3), 308–326.

Judson, E. and D. Sawada (2002) Learning from past and present: Electronic response systems in college lecture halls. *Journal of Computers in Mathematics and Science Teaching* 21(2), 167–181.

Kift, S. (2006) Using an audience response system to enhance student engagement in large group orientation: A law faculty case study. In *Audience Response Systems in Higher Education: Applications and Cases*, ed. D. A. Banks, pp. 80–95. Hershey, PA: Information Science Publishing.

King, D. B. and S. Joshi (2008) Gender differences in the use and effectiveness of personal response devices. *Journal of Science Education and Technology* 17, 544–552.

Kozma, R. B. (1991) Learning with media. *Review of Educational Research* 61(2), 179–211.

Kozma, R. B., J. Russell, T. Jones, E. Katman, N. Marx, N. Davis, and J. Wykoff (1993) Interactive multimedia and mental models in chemistry: A final report to the National Science Foundation, DUE-9150617.

Landis, C. R., A. B. Ellis, G. C. Lisensky, J. K. Lorenz, K. Meeker, and C. C. Wamser (2001) *Chemistry ConcepTests: A Pathway to Interactive Classrooms*. Upper Saddle River, NJ: Prentice Hall.

MacArthur, J. R. and L. L. Jones (2008) A review of literature reports of clickers applicable to college chemistry classrooms. *Chemistry Education Research and Practice* 9, 187–195.

Mazur, E. (1997) *Peer Instruction: A User's Manual*. Upper Saddle River, NJ: Prentice Hall.

McConnell, D. A., D. A. Steer, K. D. Owens, J. R. Knott, S. Van Horn, W. Borowski, J. Dick, A. Foos, M. Malone, H. McGrew, L. Greer, and P. J. Heaney (2006) Using conceptests to assess and improve student conceptual understanding in introductory-geoscience courses. *Journal of Geoscience Education* 54(1), 61–68.

Moore, J. L. and T. R. Rocklin (1998) The distribution of distributed cognition: Multiple interpretations and uses. *Educational Psychology Review* 10(1), 97–113.

Nakhleh, M. B., W. J. Donovan, and A. L. Parrill (2000) Evaluation of interactive technologies for chemistry web sites: Educational Materials for Organic Chemistry (EMOC). *Journal of Computers in Mathematics and Science Teaching* 19(4), 355–378.

O'Connor, V., M. Groves, and S. Minck (2006) The audience response system: A new resource in medical education. In *Audience Response Systems in Higher Education: Applications and Cases*, ed. D. A. Banks, pp. 222–247. Hershey, PA: Information Science Publishing.

Paivio, A. (1990) *Mental Representations: A Dual Coding Approach.* New York: Oxford University Press.

Pea, R. D. (1993) Practices of distributed intelligence and designs for education. In *Distributed Cognitions: Psychological and Educational Considerations*, ed. G. Salomon, pp. 47–87. Cambridge, U.K.: Cambridge University Press.

Pelton, L. F. and T. Pelton (2006) Selected and constructed response systems in mathematics classrooms. In *Audience Response Systems in Higher Education: Applications and Cases*, ed. D. A. Banks, pp. 175–186. Hershey, PA: Information Science Publishing.

Perkins, D. N. (1993) Person-plus: A distributed view of thinking and learning. In *Distributed Cognitions: Psychological and Educational Considerations*, ed. G. Salomon, pp. 88–110. Cambridge, U.K.: Cambridge University Press.

Preszler, R. W., A. Dawe, C. B. Shuster, and M. Shuster (2007) Assessment of the effects of student response systems on student learning and attitudes over a broad range of biology courses. *CBE Life Sciences Education 6*(1), 29–41.

Ruder, S. M. (2008) Depiction of curved arrow notation in organic chemistry using electronic response systems. *Abstracts of Papers of the American Chemical Society 235*, CHED-1561.

Salomon, G. (1979) *Interaction of Media, Cognition, and Learning.* San Francisco: Jossey-Bass (also published in 1994 by Routledge).

Salomon, G., ed. (1993a) *Distributed Cognitions: Psychological and Educational Considerations.* Cambridge, U.K.: Cambridge University Press.

Salomon, G. (1993b) No distribution without individual's cognitions. In *Distributed Cognitions: Psychological and Educational Considerations*, ed. G. Salomon, pp. 111–138. Cambridge, U.K.: Cambridge University Press.

Salomon, G. and R. E. Clark (1977) Re-examining the methodology of research on media and technology in education. *Review of Educational Research 47*(1), 99–120.

Sauers, A. L. and R. W. Morrison (2007) In-lecture guided inquiry for large organic chemistry classes. *Abstracts of Papers of the American Chemical Society 233*, CHED-838.

Silberberg, M. S. (2009) *Chemistry: The Molecular Nature of Matter and Change.* New York: McGraw-Hill.

Tobias, S. (1990) *They're Not Dumb, They're Different: Stalking the Second Tier.* Tucson, AZ: Research Corporation.

University of Wisconsin (2006) Chemistry conceptests. http://www.chem.wisc.edu/~concept/ (accessed April 1, 2009).

Vygotsky, L. S. (1978) *Mind in Society.* Cambridge, MA: Harvard University Press.

Woelk, K. (2008) Optimizing the use of personal response devices (clickers) in large-enrollment introductory courses. *Journal of Chemical Education 85*(10), 1400–1405.

CHAPTER 13

A Walk on the Applied Side: Developing Hypermedia for Physical Chemistry

ERIK M. EPP and GABRIELA C. WEAVER

Purdue University, West Lafayette, IN

TECHNOLOGY IN THE CLASSROOM

Out of the debate over the use of technology in education has risen the idea that technology is not an end unto itself (Cuban 2001) but a means to reach other ends more efficiently. In this chapter, we examine a situation where technology facilitated student exposure to applied research in physical chemistry and advanced instrumentation.

THE ISSUES: PHYSICAL CHEMISTRY LABORATORY

Course Issues

Physical chemistry is an upper-level course at most institutions and is generally only required for a few majors. As such, course enrollment is at least an order of magnitude below that of general chemistry, and sometimes much less. Thus, physical chemistry, and its associated laboratory component, where offered, have a small enough enrollment that it can be hard to justify the cost of specialized instrumentation such as spectrometers, lasers, computer controls, and other such equipment

Any opinions, findings, and conclusions or recommendations expressed in this material are those of the author(s) and do not necessarily reflect the views of the National Science Foundation.

that would receive little use in comparison to their high initial cost and maintenance requirements. For such situations, a technology-based solution can help to provide access to the content that would otherwise require these instruments, but at a much lower cost.

Content Issues

Physical chemistry is known for its heavy use of mathematics, which can makes the subject seem abstract. Therefore, it is no surprise that mathematics ability has been found to be a good predictor of student success in physical chemistry (House 1995; Bers 1997; Nicoll and Francisco 2001). The emphasis on mathematics, combined with a traditional focus on the early development of thermodynamics and quantum mechanics, leads to the impression that physical chemistry is a theoretical science, with only tenuous connections to "real" science. However, making connections to concepts or applications that are known to or relevant to students can be difficult or impossible if attempted through a hands-on approach. Equipment costs and availability, time limitations, and expertise can all hinder such efforts. There again, a technology-based approach can provide a solution.

Addressing the Issues

The *Physical Chemistry in Practice* DVD was developed too address these issues. It presents applications of physical chemistry to modern research and to real-world topics. It uses an easily distributed medium— the DVD disk—to showcase instrumentation not commonly available to students. The DVD uses video documentary linked to content via a hypermedia interface. The interface and its implications for learning are a focus of our research on this DVD.

THE TECHNOLOGY: HYPERMEDIA

Hypertext is "a body of written or pictorial material interconnected in such a complex way that it could not conveniently be presented or represented on paper" (Nelson 1965, p. 96). Hypermedia is a broader analogue to hypertext where the material is not necessarily restricted to textual and graphical mediums, but may contain audio and video as well interactive components. We use the term medium both to refer to hypermedia as a whole, as well as the individual classifications of content contained within.

Although the idea of hypermedia is rather old (Bush 1945; Greenberger 1964), its use in education has only been popularized recently, with the widespread adoption of the World Wide Web and hypertext markup language (HTML) technology. Its impact on students' lives outside the classroom is multifaceted and has increased students' familiarity with the technology. This has led to the desire to reexamine how students interact with hypermedia because many of the early conclusions might also be attributable to the novelty of the medium (Clark 1983; Conklin 1987).

Learning from Hypermedia

Knowledge is highly interconnected and cannot be directly transferred to the written page. Because hypermedia can store large, interconnected data structures, hypermedia has been viewed as a way for storing information in a way that parallels how people think (Bush 1945). Particularly in light of this idea, the reverse process has been intriguing—can these structures be used to aid learners in understanding material? The DVD presents video, audio, graphical, and textual information in an interface that allows users to access these as they need them.

Constructivism

Constructivism, with its tenet of the active construction of knowledge by the learner (Bodner 1986; Ferguson 2007), rules out the direct transfer of understanding from the medium to the learner. This focuses the emphasis on the active process—how the learner interacts with the material and the medium—such as hypermedia in the DVD—to build understanding. This means that the order in which a learner is presented with material affects the learning outcome. With hypermedia, the learners can choose the order and pace of accessing information, giving them great flexibility, and allows them to work in the way that they learn best.

Cognitive Load Theory

This flexibility is not without a price, however, as learners need to keep track of where they are and pace themselves. To address this, we examine the situation from the psychological perspective of cognitive load theory. Cognitive load theory is based on the premise that people have a limited working memory capacity and that information must flow through before it can be stored in long-term memory (Sweller 2005). The basis for this is Miller's work, which indicated most people had a working memory capacity of 7 ± 2 items (Miller 1956).

From this foundation, several frameworks for function have been developed (Baddeley 1992; Johnstone 1997; Mayer and Moreno 1998). Each shares the commonality of having the working memory as an intermediate storage location before long-term memory. The storage and processing of information in working memory is a *prerequisite* for constructivist learning. Before a learner can construct knowledge, it must pass through working memory, a limited resource.

For hypermedia, learners have a cognitive load related to keeping track of their location within the content and how they interact with the interface. This is in addition to the cognitive load of the content itself. In contrast, textbooks are linear in order and students are familiar enough with the "interface" that little cognitive load is generated. The hypermedia interface of the *Physical Chemistry in Practice* DVD has been studied and refined to provide students with blocks of information that can be dealt with at a single sitting. Also, there are numerous ways for users to track their place in the content, through nested menus.

THE *PHYSICAL CHEMISTRY IN PRACTICE* DVD

The DVD has several formats of information and is divided into a set of different content topics. We will describe the content topics first, followed by the interface and layout of information in the DVD.

Content Modules

The content consists of an introduction to the DVD and eight topical modules. Each module focuses on a topic in applied, physical chemistry research and is presented in a variety of media. The topics were selected to have very little overlap, and thus a wide breadth of concepts is covered. The settings are also varied, including industry, medical, and academic research settings. An overview of the topics is listed below:

- *Atomic Force Microscopy (AFM)* This module examines the techniques behind AFM and shows its applications in studying surfaces, particularly silicon surfaces, such as those that might be used in semiconductors and computer chips.
- *Electronic Structure of Vitamin B_{12} Corrinoids* Vitamin B_{12} is a complex molecule and its activation pathway is not fully understood. Research using UV-vis, resonance Raman, circular dichro-

ism, and magnetic circular dichroism spectroscopies along with density functional theory seeks to reconcile information from each of these techniques to better understand vitamin B_{12}'s electronic structure.

- *Bose–Einstein Condensates*   Bose–Einstein condensates are the coldest substances made by humankind. This module explores how matter can be cooled to such a great extent, and the applications of coherent matter.
- *Single-Molecule Manipulation of DNA*   Optical traps and micropipettes are used to hold and to stretch single molecules of DNA, showing how the folded strand behaves as an entropic spring.
- *Solid-Acid Electrolytes and Hydrogen Fuel Cells*   Fuel cells are an alternative method of generating energy; however, advances need to be made in the electrolytes to improve the efficiency. This module examines one new class of electrolytes—solid acids.
- *Thin-Film Polymer Kinetics*   This module explores how kinetics applies to the thin films used in the photolithography of microprocessors. By limiting diffusion in the solid state, even smaller features can be created, making smaller, more efficient chips.
- *Nuclear Magnetic Resonance (NMR) and Magnetic Resonance Imaging (MRI)*   The use of imaging in the medical profession actually takes its roots from NMR work in chemistry. This module shows how the techniques of NMR were expanded to allow 3-D imaging.
- *Surface-Enhanced Raman Spectroscopy (SERS)*   SERS is a technique for increasing the sensitivity of a Raman instrument. This module explores its application to the detection of chemicals indicative of brain injury.

Elements of Design

Taking advantage of new technology by NetBlender, *Physical Chemistry in Practice* is an enhanced DVD that allows for standard video playback in a DVD player but also has an enhanced interface when used with a Microsoft Windows computer (Fig. 13.1). While any DVD player can play the video contained within, the enhanced interface provides access to the same video while adding flexibility of access and supporting content in other formats. The content of the enhanced interface is summarized in Table 13.1, and the major content sections are described in detail below.

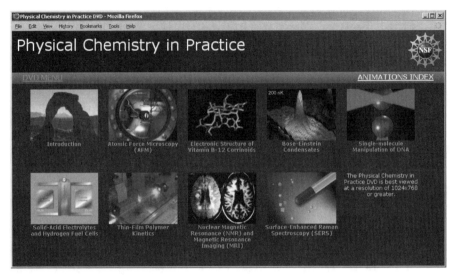

FIGURE 13.1   Main screen of the DVD in the enhanced interface.

TABLE 13.1   Enhanced Interface Content

Content Type	Description
Videos	Documentary movies about the topics, showing interviews with the scientists and footage from their laboratories
Theory	Detailed descriptions of the concepts mentioned in the video, going into greater detail than the videos provide and connecting to standard "textbook" concepts
Animations	3-D and 2-D animations created specifically to highlight or to demonstrate concepts in the videos
Definitions	Key terms and concepts
Problems	Questions for students over the material covered
Transcript	A verbatim transcript of video
References	External materials that may be of use in understanding the topic including research literature or URLs
Credits	List of people involved in the production of the module

Videos

The video section consists of DVD-format documentary-style videos. For each module, there are interviews with the principal investigator and with associated researchers discussing the context and purpose of their work. Laboratory procedures and equipment are demonstrated and explained. Computer-generated animations show molecular-level representations and diagram processes.

Each module has 13–19 video chapters for a total length of 30–45 minutes with individual chapters of 1–6 minutes in length each. The short chapter lengths were selected to keep the students' attention and to make it easy to switch mediums if desired. The videos are stored in DVD standard MPEG-2 format and are accessible in a DVD player or through the enhanced interface on a computer. The animations integrated into the videos are also available individually (described below).

Animations

Animations are used in the videos to show submicroscopic representations, internal workings of instrumentation, and other important processes. These are integrated into the videos (above) at appropriate times to illustrate what is being discussed.

The computer-generated animations are also available separately from the video, as stand-alone clips. The animation clips are accompanied by the narration audio that is present when played with the video. When viewed by students, this allows them to review the animation material at their own pace. If used by instructors, they can mute the audio in order to provide their own explanations.

Theory

The theory sections consist of explanations of concepts in or related to the research discussed in the module. They are structured to parallel the content of the videos, allowing easy access for students while watching the videos. The format of the information is similar to a textbook, with written explanations accompanied by appropriate schematics, figures, and equations. The theory pages are stored as HTML in an XML repository, accessible only through the enhanced interface.

Definitions

Since the topics discussed in the DVD extend beyond the normal subjects covered in most physical chemistry courses, a glossary of terms is provided. Additionally, terms are associated with the video chapters in which they appear, allowing students easy access to the definitions as they encounter them. The definitions pages are stored as HTML in an XML repository, accessible only through the enhanced interface.

Problems

Questions are included to give students a chance to apply their learned knowledge to realistic problems from the research. Where possible, the problems are based on actual research data provided by the principal investigator whose work is being discussed. For instance, in the AFM module, actual AFM images are available for students to examine island formation under epitaxial growth. However, this is not always possible, so problems tying familiar physical chemistry concepts to the research are used in some cases. The problems are stored as HTML in an XML repository, accessible only through the enhanced interface. Although interactive questions with immediate feedback were planned, we were unable to adequately develop them, so solutions and feedback on the questions would need to be addressed by the instructor.

Transcript

A verbatim transcript of everything that is said in the audio/video is provided to the students. This also proves useful for students to find the spelling of unfamiliar terms or to clarify speech that may be difficult to understand. This also makes the videos' contents accessible to the hearing impaired, since we were unable to provide closed captioning for the videos. The transcript is stored as HTML in an XML repository, accessible only through the enhanced interface.

References

References to the primary literature are included to allow students to further research the topic. This also provides the instructor with easily accessible resources to expand on the material. The references are stored as HTML in an XML repository, accessible only through the enhanced interface.

INTERFACE

Chapter Mode

This mode was added after the first implementations, in response to student feedback. Despite the research literature to the contrary (Sweller et al. 1998), students wanted the ability to view the video with associated text (theory, definitions, and transcript) alongside. The chapter mode allows users to have the video playing on one-half of the screen and text on the other (Fig. 13.2).

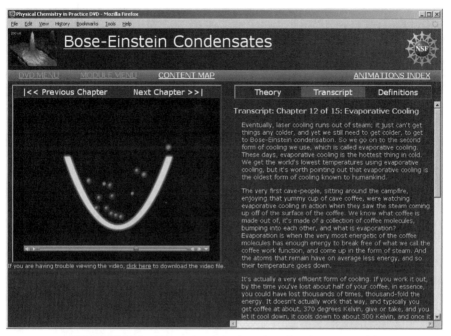

FIGURE 13.2   Example of the interface for chapter mode.

DEVELOPMENT PROCESS

Selection of Topics

Topics were selected by examining the research literature in physical chemistry for projects that had a relevant connection to students' lives or future careers while also considering if the research would lend itself to being filmed. As topic candidates were found and compiled, the list was matched to topics traditionally covered in the physical chemistry course to allow the DVD to be easily tied into an existing course. To choose the final set of topics, we decided to pick the broadest range of topics to provide a variety for instructors and to have ties to as many topics covered in physical chemistry courses as feasible.

Video Editing—First Pass

Once principal filming was done, the video segments were examined for clarity of dialogue, and then ordered for content structure and flow, such as determining where the chapter breaks should go. The segments were then transcribed and marked up to indicate where the cuts should go. Some portions were marked to be covered with animations or "B-roll" film (location/instrumentation shots). This transcript was then given to the video editing staff to assemble. Once assembled, a rough cut of the video was generated to verify that the content was built appropriately and was understandable.

Computer-Generated Animations

Once the rough cut was assembled, work began on the animations, since the timing of the video is important in generating the animations. A content specialist would generate the list of animations and create storyboards for each. The content specialist would then work with a computer animator to generate the necessary graphics. This process was an iterative one with several drafts of animations being prepared along the way. It was important for the content specialist to maintain close contact with the animator to prevent work in unfruitful directions.

Video Editing—Second Pass

With the animations complete, the video editing staff would compile a final edit that included all animations, B-roll, and chapter cuts in the proper places. This version would then be proofed by the content development team and the principal investigator.

Content Writing

Concurrent to the video production was the writing of the text that would accompany it. The transcript was generated from the script, with "uh"s and similar items being removed. The theory sections were assembled by drawing from original research papers and physical chemistry textbooks. From the theory and transcript, terms that students might have been unfamiliar with were selected for the definition sections.

IMPORTANCE OF THE INTERFACE

There is a cognitive load associated with each medium, which affects the level of cognitive effort required to access the information. While older technologies like text on paper have almost no load due to the learner's familiarity with them, technologies with which a learner is not as familiar require cognitive resources when being accessed.

Although the hypermedia interface allows great freedom to students, this also places a cognitive burden on students to keep track of where they are and where to go next. First noted by Conklin (1987), the "lost in hyperspace" effect has led to the use of advance organizers to try to overcome this (Brinkerhoff et al. 2001). An additional approach is to include a suggested path that sequential learners find helpful and does not detract from the learning process for global learners (Dünser and Jirasko 2001). Sequential learners prefer material in an ordered arrangement, while global learners are comfortable without a guide. Both suggested paths and advance organizers were used in the implementation of the *Physical Chemistry in Practice* DVD.

IMPLEMENTATION

For this study, a physical chemistry laboratory course at a large, Midwestern university was selected. The use of a three-hour laboratory period for student interaction with the DVD ensured sufficient time for students to explore, and gave us a naturalistic setting in which their interactions could be observed. The week before the laboratory, students were given a handout with a description of the module and a list of the major content divisions. Students were permitted to take notes and to use these notes on their assignment, which consisted of a series of questions over the material.

While the overall development used a Windows application programming interface, incompatibilities exist across platforms (Microsoft Windows, Apple OS X). For this reason, we opted to use a more standardized interface using HTML that would be more accessible, at the cost of a few features (pairing text to certain times in the videos and full-screen video). Aside from these two features, the HTML interface is identical to the enhanced interface, using the same color scheme and layout and containing the same content.

DATA COLLECTION

Our goal was to observe the use of the DVD by students as they learned from it in preparation for their assignment. Keeping with the desire to study students' usage in a naturalistic setting, the use of video cameras was ruled out as too intrusive. Due to the encrypted nature of DVD video, standard screen capture software was unusable. Therefore, the low-tech solution of VCR recordings was used to capture video output directly from the computer. For this option, VCRs were connected to a secondary video output on the computer via an adapter cable.

DATA PROCESSING

The video tapes were transcribed by reviewing them and, at each transition from one page to another, by recording the time on the tape and the location on the DVD. This provided a navigation log of the order in which students viewed information and how much time was spent on each page.

These data were then analyzed by plotting the logs as "maps" of each student's path through the material (Fig. 13.3). The horizontal axis marks the passage of time, while the vertical axis gives the location in the DVD (arbitrarily assigned). Each x, y point is marked by a circle. The area of the circle represents the amount of time spent on that particular page. Additionally, the data were analyzed in bulk, looking for overall trends in usage.

RESULTS—INDIVIDUALS

When we examine the paths that students take, we see a variety of behaviors. Figure 13.3 shows one sort of behavior we were expecting, a student viewing a variety of media. As the different patterns in the

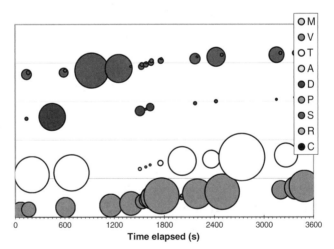

FIGURE 13.3   Mixed-media chart. M = main screens; V = video; T = theory; A = animations; D = definitions; P = problems; S = transcript; R = references; C = credits.

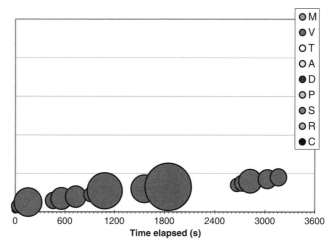

FIGURE 13.4   Video only (M = main screens, V = video, T = theory, A = animations, D = definitions, P = problems, S = transcript, R = references, C = credits).

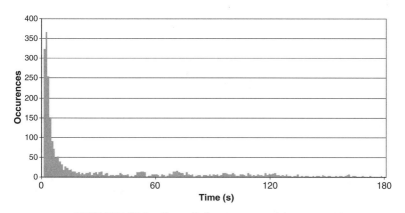

FIGURE 13.5   Overall time-per-page histogram.

circles show, the student switched between different types of content several times.

Figure 13.4 shows another student behavior; this student primarily progressed through the video section in order with no visits to other types of content.

RESULTS—OVERALL TRENDS

Switching to analysis of the entire set of navigation logs, we show a histogram of the amount of time spent on a given page (Fig. 13.5). Since

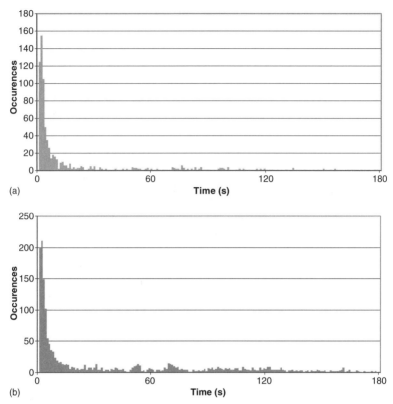

FIGURE 13.6   (a) Menu time-per-page histogram. (b) Non-menu time-per-page histogram.

there is such a large skew toward small amounts of time (<10 s), too short to actually process the information on a page, we split out time spent on the menu system (Fig. 13.6a) from the time spent on content (Fig. 13.6b). Interestingly, we see the same sort of pattern in both cases—a large number of instances of very little time spent.

From this analysis, we see that students do take advantage of the flexibility afforded to them. Different students navigate through the material in drastically different ways. From a cognitive load perspective, this shows that the intrinsic load imposed by the medium does not seem to be inhibiting student behavior.

The time-per-page analysis shows that many students spend far less time on content than one would think. With 47.1% of interactions being 10 s or less on pages containing content, we are unsure of what students are doing in that brief amount of time. The choices that cognitive load theory gives us are that they either are not processing the information presented or that they are only processing a small fraction of it. The first could be indicative of Conklin's (1987) "lost in hyperspace" problem—that the students are attempting to figure out where they are. The latter would indicate that students are using some technique to selectively pick out small amounts of information to process.

Although the data give us a good characterization on how students spend their time when learning from hypermedia, they also raise many questions on why students make the choices they do when navigating through hypermedia. To get a better understanding of the choices students make, further work using think-aloud protocols is under way in order to elucidate student decision-making processes when navigating hypermedia.

ACKNOWLEDGMENTS

The development work was funded in part by National Science Foundation (NSF) CCLI-EMD #0127541. The development of the *Physical Chemistry in Practice* DVD would not have been possible without the efforts by the following: Peggy O'Neill-Jones, Beatriz Cisneros, Joseph Farris, and Justin Heisler. The authors would also like to thank the researchers who graciously allowed us into their laboratories: William Bradley, Thomas C. Brunold, Carlos Bustamante, Eric Cornell, Gerard Coté, Sossina Haile, Bill Hinsberg, Frances Houle, Stephen Leone, John D. Roberts, and Richard Schwenz.

REFERENCES

Baddeley, A. (1992) Working memory. *Science 255*(5044), 556–559.

Bers, T. (1997) Self assessments, academic skills, and student achievement. *Journal of Applied Research in the Community College 4*(2), 101–117.

Bodner, G. M. (1986) Constructivism: A theory of knowledge. *Journal of Chemical Education 63*(10), 873–878.

Brinkerhoff, J. D., J. D. Klein, and C. M. Koroghlanian (2001) Effects of overviews and computer experience on learning from hypertext. *Journal of Educational Computing Research 25*(4), 427–440.

Bush, V. (1945) As we may think. *Atlantic Monthly*. http://www.theatlantic.com/doc/194507/bush (accessed February 25, 2009).

Clark, R. E. (1983) Reconsidering research on learning from media. *Review of Educational Research 53*(4), 445–459.

Conklin, J. (1987) Hypertext: An Introduction and Survey. *Computer 20*(9), 17–41.

Cuban, L. (2001) *Oversold & Underused: Computers in the Classroom.* Cambridge, MA: Harvard University Press.

Dünser, A. and M. Jirasko (2001) Interaction of hypertext forms and global versus sequential learning styles. *Journal of Educational Computing Research 32*(1), 79–91.

Ferguson, R. L. (2007) Constructivism and social constructivism. In *Theoretical Frameworks for Research in Chemistry/Science Education*, eds. G. M. Bodner and M. Orgill, pp. 27–49. Upper Saddle River, NJ: Prentice Hall.

Greenberger, M. (1964) The computers of tomorrow. *Atlantic Monthly*. http://www.theatlantic.com/unbound/flashbks/computer/greenbf.htm (accessed February 26, 2009).

House, J. D. (1995) Noncognitive predictors of achievement in introductory college chemistry. *Research in Higher Education 36*, 473–490.

Johnstone, A. H. (1997) Chemistry teaching—Science or alchemy? *Journal of Chemical Education 74*(3), 262–268.

Mayer, R. E. and R. Moreno (1998) A split-attention effect in multimedia learning: Evidence for dual processing systems in working memory. *Journal of Educational Psychology 90*(2), 312–320.

Miller, G. A. (1956) The magical number seven, plus or minus two: Some limits on our capacity for processing information. *Psychological Review 63*(2), 81–97.

Nelson, T. H. (1965) Complex information processing: A file structure for the complex, the changing and the indeterminate. In *Proceedings of the 1965 20th National Conference*, ed. L. Winner, pp. 84–100. Cleveland, OH: ACM.

Nicoll, G. and J. S. Francisco (2001) An investigation of the factors influencing student performance in physical chemistry. *Journal of Chemical Education 78*(1), 99–102.

Sweller, J. (2005) Implications of cognitive load theory for multimedia learning. In *The Cambridge Handbook of Multimedia Learning*, ed. R. E. Mayer, pp. 19–30. Cambridge, UK: Cambridge University Press.

Sweller, J., J. J. G. van Merriënboer, and F. G. W. C. Paas (1998) Cognitive architecture and instructional design. *Educational Psychology Review 10*(3), 251–296.

CHAPTER 14

Effective Use of Games and Puzzles in the Chemistry Classroom

THOMAS D. CRUTE III

Augusta State University, Augusta, GA

Syndicated columnist and psychologist John Rosemond strongly advocates that successful child rearing requires a bit of humor and fun in the family (Rosemond 2008). The same may be said of the successful classroom. Humor and fun can be provided by using classroom games to help engage students. As it is generally recognized that students desire occasional "breaks" in the classroom (Smith 2006), a game activity can provide this relief and reduce student stress. However, despite these benefits, games must be purposefully chosen with a goal of achieving real learning. Learning is serious business, but learning is facilitated with successful strategies to engage and to interest students. An interesting or fun activity has a high likelihood of engaging students and curriculum-based games and puzzles accomplish this.

Besides fun, to be effective, game activities must serve a legitimate classroom purpose and must lead to the fulfillment of educating students in chemistry. Whether the game introduces new material or provides a forum for the refinement and reinforcement of concepts already taught, game activities should be approached as educational tools. For example, a game may be a review of specific topics that were otherwise already taught in the classroom either to develop new interconnections, to uncover new nuances, to provide additional experience with working through the concepts, or as a motivational or self-check tool for a student to assess readiness for a traditional test. As preparation for a test, a game may bring together many topics that were not previously considered at once. Alternatively, the game may review material learned independently from a reading or out-of-class assignment. A game may be used in and of itself as an assessment tool instead of a more traditional quiz or test. Sometimes, the game can be used to teach a concept or skill independent of more traditional instruction. There are also chemistry games that provide effective analogies by having game activities that mimic molecular behavior (*vida infra*).

PRACTICAL CONSIDERATIONS

With so many worthwhile learning activities competing for limited class time, any game activity should pass muster on the question of whether completing it was a valuable use of the time and effort. A game should be used as a learning or review tool and not merely as a time killer or as a recess period.

Other considerations should include whether the particular game material is appropriate for the course or the specific goals of the unit under study. For instance, a word search (Helser 2008) can provide a forum for students to work with specific vocabulary. Such an activity may be very effective at the beginning of a unit as a way to prompt students as to which vocabulary will be necessary to learn. However, if the learning goal is to be able to clearly define the words or to use the vocabulary properly in context, then such an activity would not provide the correct focus in this example. An alternative that better fits this goal might be a crossword puzzle (Most 1993), an acrostic (Swain 2006), or a Jeopardy-type game (Keck 2000) where the activity specifically requires matching vocabulary to definitions.

When incorporating games into the classroom, one should carefully evaluate the preparation necessary for a particular game activity. From an instructor standpoint, some games require a significant amount of preparation. Such an investment in time, effort, and materials would be needed to be balanced against the value. This would be especially true of board game activities (Russell 1999a) whereby significant preparation is necessary and only a handful of students can participate at a time. While one could prepare several sets of materials so everyone can be involved at once, a more practical solution is to allow the materials to be used by different groups at different times over the course of several class periods. Other activities would then be necessary for those not playing that particular game.

The instructor must carefully weigh the amount of student preparation necessary to carry out a game activity. There is more to the preparation for a classroom game than the creation or gathering of materials by the instructor. Some games, especially ones with complex or unfamiliar rules, require a certain degree of investment by the students in order to effectively participate. While many valuable games are simple and intuitive, others may have a very high level of complexity. For instance, a game on nuclear synthesis and decay requires a rather lengthy set of rules (Olbris and Herzfeld 1999), but the benefit is a mimicking of results that cannot easily be performed in the laboratory along with a self-discovery of new concepts and trends. One may reasonably conclude that the considerable preparation necessary to train the students on how to play this game is well worth the effort in the benefits of the game. Even if the instructor decides that the worth of a game activity warrants a significant investment of time and effort, the students must also buy in to the worth of the activity if they are to be sufficiently engaged.

Another important consideration has less to do with the activity itself and more to do with setting the appropriate tone in the classroom given the characteristics of the students involved. Many games are competitive in nature either as a self-challenge or as a head-to-head competition with other individual students or teams of students. On the one hand, a healthy level of competition is beneficial. Students participating in a competitive situation will usually take the game seriously and therefore are more likely to benefit from the intended learning outcomes. Thus, a competitive situation may bring out the best in a student as the pride of winning motivates the individual. Especially in team-based games, individuals may be motivated so as to not be the weak link on the team. Taken to an extreme, however, competition may be counterproductive. Emotions may run high and overzealous students may take the competition to an unhealthy level. Differences in abilities of stronger and weaker students that may have been more private in individual assessment activities may be highlighted or even magnified.

ADAPTATIONS OF NONGAME ACTIVITIES INTO GAME ACTIVITIES

Incorporation of games into your classroom need not be difficult. One of the easiest methods is to adapt an existing exercise in your curriculum into a game. For instance, a practice worksheet may be modified into a game to add interest to the activity. In one published example, a sheet of alkene reactions had possible product structures listed that corresponded to a letter of the alphabet. When all problems were solved, the letters provided the answer to a riddle that was posed (Bertolini and Tran 2006).

Practicing some chemistry skills as a group may be quite beneficial to learning and a game platform may facilitate the practice. The learning of molecular shapes and valence shell electron pair repulsion (VSEPR) theory frequently benefits from hands-on manipulation. Cooperative learning of these concepts as part of a game (Myers 2003) has teams of students using simple materials to assemble models of the compounds being studied with the additional challenge of competing head to head with other teams. This activity is best suited for review of concepts rather than for initial learning.

Another strategy for converting a nongame activity into a game is to have students collect the results of chemical information gathering and to assemble it into a game. For instance, in one published game, students are instructed to collect information on the biological significance of their assigned elements (Sevcik et al. 2008b). This preparation allows the students to categorize the biological necessity or the toxicity of the assigned element and thus to color-code it. Once all the color-coded game pieces constructed by the students are assembled, students toss Velcro-covered balls at the color-coded periodic table to score points based on which category of element they hit.

Yet another example of creating a game from scratch is to turn curriculum material into a "Twenty Questions"-type game. In this type of game, students ask yes or no questions about an instrument, element, famous scientist, and so on, to gather enough information to be able to identify it correctly in a group setting (Koether 2003). Variations of this include a deduction game whereby each student is assigned an identity (such as an element) that is unknown to them but is visible to the other class members on an identity card. The students then "mingle" and ask yes or no questions about their own identity until it is determined. Another variation goes beyond yes or no questions and focuses instead on synonyms while avoiding taboo words as clues for guessing the keyword are given (Capps 2008). These sorts of games can be readily customized to a variety of topics that may be present in the curriculum.

A similar concept to the Twenty Questions-type game is a game that asks students to match items that are related to each other. For instance, beginning students may benefit from matching the name of laboratory equipment to the equipment itself. Making a game out of this is readily achieved (Greengold 2005) and will likely be better received by students than a list to memorize. A similar concept is an adaptation of a "concentration" game. In the traditional concentration game, two instances of the same object or idea are present in hidden form on a grid, and the challenge is to remember the location of the hidden matched items. In a learning environment, instead of using the exact match, a directly related word or object could be used. Published examples include matching an important vocabulary word with its definition (Nowosielski 2007) or matching element names to symbols (Sevcik et al. 2008a).

Some chemistry concepts require learning or developing stepwise processes. Such activities can be reinforced by constructing a game that requires the players to assemble process steps in the correct order. This could be used for general problem-solving algorithms (Bartholow 2006) or for specific stepwise chemical processes such as reaction mechanisms (Erdik 2005). Multistep synthesis can be adapted into a cooperative team event as well (Crute 1992) by having each reaction of the sequence provided in relay form by each member of a team.

ADAPTATION OF TRADITIONAL GAMES

The examples in the section demonstrate that a wide variety of classroom activities can be readily transformed into a simple game. An alternative to inventing a game that adapts to existing curricular materials is to adapt a well-known game to chemistry. Perhaps the most common example uses the basics of a Jeopardy-type game (Scarpetti 1991; Deavor 1996; Keck 2000). As a tool for reviewing concepts, the Jeopardy format allows the mixing of a wide variety of concepts into a single game by virtue of having different categories. While one may prepare such a game using individual cards or sheets of paper with point values, answers, and questions, the availability of computer technology can simplify matters greatly by using tools such as hyperlinked images arranged on a playing grid. Templates for setting up a game in this manner are available as HTML (Grabowski and Price 2003) or PowerPoint slides (Oppertshauser 2008).

Logical deduction games are commonly used to develop logical thinking (Gottstein 1998). A typical game would use process of elimination and other simple logical thinking when offered limited individual bits of information (e.g., five friends gave birthday gifts; John gave a book; Mary was the neighbor of someone who gave a book; and two partygoers were twins and only one of them gave a book) The problem solver then uses the clues to determine which person gave which gift in this example. A chemistry variation bases the logic on chemistry information and behavior (Peris 2007).

The familiar game of Bingo can be adapted to chemistry as well. In a published version of this game (Crute and Myers 2000), students practice chemical nomenclature and use the pieces of the correct names to match name fragments on the Bingo card. Naming the substance incorrectly would lead to improperly filling out the Bingo card. Thus, students must be both accurate to mark the correct spots and fast to beat other students in a head-to-head competition.

Crossword puzzles and Sudoku puzzles may be found in nearly any newspaper and have been applied to chemistry as well. Crossword puzzles may be developed to teach or to review a variety of material that can be summarized as single word or several word answers based on the clue given. As such, the chemistry version mirrors the traditional versions.

Traditional Sudoku puzzles, on the other hand, use numbers to test the logic skills of the solver. As effective chemistry classroom puzzles, the numbers have been replaced by chemical symbols or pictures. The value of completing such an exercise could be found in using nonidentical, yet matched, symbols instead of numbers. Students must then learn to recognize which symbols are matched. Examples might include names and symbols of elements, names and structures of organic functional groups, and names and/or abbreviations and/or structures of amino acids. Instructions for developing your own chemistry Sudoku (Crute and Myers 2007) as well as ready-to-use puzzles (Perez and Lamoureux 2007; Saecker 2007; Welsh 2007) have been published.

CUSTOM GAMES

The dynamic nature of molecules can be troublesome for students. To aid in understanding molecular behavior, a number of games mimic the activities that chemical compounds may undergo. To duplicate chemical equilibrium, students throw dice in one published game (Edmonson and Lewis 1999). By participating in the equilibrium firsthand, students may better comprehend the system.

In a similar manner, game activities provide the data for the amount or type of change for other processes. There are a number of games that duplicate the statistical nature of some chemistry concepts by using statistical methods in the game activity. "Throwing" rock-paper-scissors to determine how to exchange money drives a Boltzmann game and a study of entropy (Hanson and Michalek 2006). In a chromatography game, coins are used to represent individual molecules of a mixture. During the play of the game, the equilibrium between interaction with the mobile and stationary phases is clearly seen as the source of chromatographic separation (Samide 2008). In another game, the statistical outcome of a game simulates kinetics to help bridge the gap between kinetic data and molecular behavior (Harsch 1984). Rate theory has also been simulated with games (Olbris and Herzfeld 2002). One could create additional games by using these sorts of statistical tools (dice, cards, etc.) or by modifying existing games with more convenient tools.

Some games serve to duplicate the practical nature of chemistry rather than the behavior of molecules. In a peptide sequencing game, students must make wise choices as they try to sequence a protein by choosing which experiments and in which order they wish to have results from (Lemley 1989).

The nature of chemistry also allows the development of games that have no peer outside of chemistry. Specialized games can be developed to teach specific topics such as organic mechanistic steps (Erdik 2005) or regiochemistry of aromatic substitution reactions (Zanger et al. 1993). Outside of the "do-it-yourself" games discussed to this point, there are computer games and commercially available chemistry games such as board and card games. A review article on chemistry games that includes contact information for the manufacturer of a number of commercially available, ready-made chemistry games is available (Russell 1999b).

INCORPORATION OF SUDOKU IN THE CLASSROOM— AN EXAMPLE

To illustrate specific considerations that may be required for implementing a game or a puzzle, a case study of design and assignment of a puzzle will be examined. First and foremost, one must identify the learning goals and choose the activity with an eye toward meeting those goals.

For our example, let us consider the arduous task of learning common amino acids not only by structure but also by both the three- and one-letter abbreviations. Furthermore, we want students to distinguish between acidic and basic side chains. Such a task is common in biochemistry either in introductory or upper-level course offerings. Frequently, the instructor will make the assignment to memorize the list and leave the students to do it (or not). Based on Herron's principle of minimum effort (Herron 1996), students will need to recognize the value of the learning goal to devote effort toward it. On the one hand, making the task more complicated by assigning it as a puzzle is contrary to this principle; however, Herron also reminds us that students will choose pleasant experiences over unpleasant ones. He also assures us that people find "different things interesting." While the development of a Sudoku puzzle to assist with this learning goal represents an increase in complexity, the puzzle aspect may make the task more enjoyable, certainly different from rote memorization of a list, and perhaps more effective by forcing attention to certain details.

In a typical Sudoku puzzle described earlier in this chapter, the grid allows for a total of nine distinct symbols. In most versions, these are the numbers 1 through 9. We will make it nine different amino acids. Of course, most instructors will wish to have students memorize at least 20 amino acids and may dismiss Sudoku as too limited for the task. Conversely, breaking this large task into smaller tasks is an advantage for learning in that the task becomes more manageable for the student. By carefully selecting which nine amino acids are to be learned, such as all the common acidic and basic side chain amino acids, we can assist the student with building the knowledge in a meaningful way. Additional amino acids that have different properties (i.e. not acidic or basic side chains) can be subsequently assigned. Breaking the task into smaller pieces is advantageous as the student will master some material earlier and then add related, yet different material to it later. If the later amino acids have only neutral side chains, students are less likely to classify them incorrectly when working with all the assigned amino acids later.

For our example, we will choose among the 20 most common amino acids first according to classification of the side chain. There are three amino acids with basic side chains: histidine/His/H, lysine/Lys/K, and arginine/Arg/R. There are two amino acids with acidic side chains: aspartic acid/Asp/D and glutamic acid/Glu/E. One of our goals is to have students recognize the acidic and basic nature of side chains, so selecting these can allow us to pursue that goal. We will also select the amide side chains that are derivatives of aspartic acid and glutamic acid: asparagine/Asn/N and glutamine/Gln/Q. Solving the puzzle will allow students to see the similarity between these derivatives, yet force students to attend to the difference. We must choose two more amino acids and, among many that may be chosen, we will use tryptophan/Trp/W and tyrosine/Tyr/Y based on some weakly compelling reasons. First of all, the one-letter abbreviations have little relationship to the name and will be among the more difficult ones to learn. If these are learned now, then most of the remaining one-letter abbreviations will be easy to learn (glycine = G, valine = V, leucine = L, etc.). Second, the phenolic group in tyrosine has a greater acidity than most of the

		2	1		6	3		
8		4			5			6
1	7		2	3				5
		9		7				2
	1		9		2		7	
3				8		1		
5				2	3		9	8
2			6			5		4
		3	8		9	6		

FIGURE 14.1 Sudoku puzzle from a newspaper or website classified as "very easy."

remaining amino acids and the other acidic side chains are already included. Tryptophan has a nitrogen-containing side chain just as others that are in our puzzle. We want students to pay attention to the differences among superficially similar compounds.

As a starting point for our puzzle, we will select an existing numerical-based puzzle that requires only the simplest logic to solve. Thus, from a newspaper or website, we can select a puzzle labeled "very easy" such as the one shown in Fig. 14.1.

In the chemistry version of this puzzle, we will substitute a particular amino acid for a particular number. Some skill with solving these puzzles is helpful here in that our recognition of the best starting places will allow us to help students distinguish certain features of the amino acids we have chosen. The easiest puzzles can usually be solved using only a "crosshatching" technique (Yates 2005). Using a crosshatching strategy on our example puzzle, we find that all instances for any of 2, 3, or 7 can be filled in without any uncertainty. Thus, if we assigned a particular amino acid for the 2's in this puzzle, then we could suggest by category that it would be a good starting place to ease the problem solving of the puzzle. In contrast, if one attempted to start filling in the instances of the 4's, one would find that there are too many blanks for which 4 could be appropriate, and thus, ambiguity exists.

9	5	2	1	4	6	3	8	7
8	3	4	7	9	5	2	1	6
1	7	6	2	3	8	9	4	5
6	8	9	3	7	1	4	5	2
4	1	5	9	6	2	8	7	3
3	2	7	5	8	4	1	6	9
5	6	1	4	2	3	7	9	8
2	9	8	6	1	7	5	3	4
7	4	3	8	5	9	6	2	1

FIGURE 14.2 Numerical Sudoku puzzle solved.

Recognizing 2, 3, and 7 as good starting places, we will assign the three basic amino acids to these values and then offer a suggestion to the student that each of the basic amino acids (lysine, arginine, and histidine, respectively) should be filled in first to facilitate solving the puzzle. Once all instances of 2, 3, and 7 are present, there are fewer blank spots available. Further solving shows that numbers 1 and 5 can now be filled in unambiguously using the crosshatching technique, even though initially starting with 1 and 5 would have been difficult. Thus, we can assign our two most acidic amino acids to the values 1 and 5 (aspartic acid and glutamic acid, respectively) and suggest to the students that after completing the three basic amino acids, they should then fill in all instances of the two acidic amino acids. Relatively few blanks remain at this point and one can now unambiguously fill in the instances of 8, 9, 4, and 6. The remaining amino acids will be assigned to these numbers: asparagine, glutamine, tyrosine, and tryptophan. The numerical version of the puzzle has the solution shown in Fig. 14.2.

The initially given numbers, now assigned to particular amino acids, will next be converted to a variety of representations for those amino acids. For some instances, the full name will be given, while others will have the structure, and still others will have either the three- or one-letter abbreviation. This way, the student must make full associations

		Lys	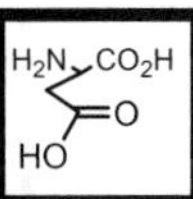 H₂N–CO₂H (=O, HO)		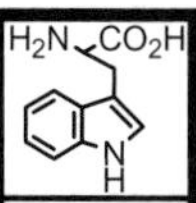 H₂N–CO₂H (indole)	Arginine		
N		Y			Glu			Tryptophan
Asp	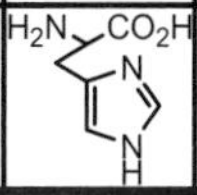 H₂N–CO₂H (imidazole)		Lysine	Arg				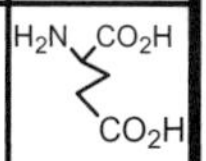 H₂N–CO₂H (CO₂H)
	H₂N–CO₂H (O, NH₂)			His				H₂N–CO₂H (CH₂)₄ NH₂
	D		Gln		Lys		H	
H₂N–CO₂H (NH, N–NH₂)				Asn	H₂N–CO₂H (CO₂H)			
H₂N–CO₂H (O, OH)			H₂N–CO₂H (NH₂)	Arg		Q	H₂N–CO₂H (=O, H₂N)
K			W			E		Tyr
		R	N		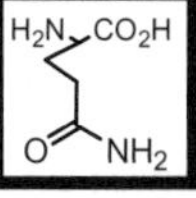 H₂N–CO₂H (O, NH₂)	Trp		

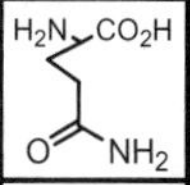

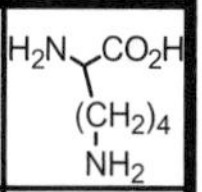

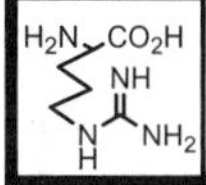

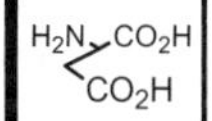

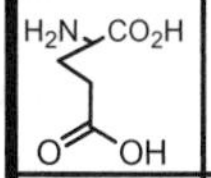

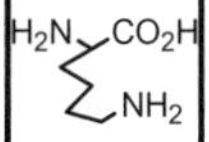

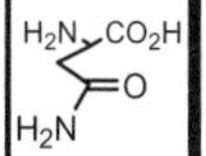

Fill in each block such that every row, every column, and every bold section contains one each of every amino acid given alphabetically here:

arginine, asparagine, aspartic acid, glutamic acid, glutamine, histidine, lysine, proline, tryptophan

The amino acid may be represented by structure, name, three-letter abbreviation, or one letter abbreviation. This particular puzzle is most easily solved by filling in all nine instances of each of the basic side chain amino acids first, then each of the acidic side chain amino acids, followed by the others. For simplicity, the amino acid structures are not represented in their zwitterionic form.

FIGURE 14.3 Amino acid Sudoku puzzle.

between the names, structures, and abbreviations. Once substitutions are made, the puzzle may look like Fig. 14.3. Note the instructions concerning which amino acids are present, the suggestion of tackling basic ones first followed by acidic ones, and a comment concerning the structures not being shown in the zwitterionic form.

Since each amino acid is represented in as many as four different symbolic ways, the answers by students will likely be quite varied. The nature of sudoku puzzles is such that there is rarely only a single error

Gln	Glu	Lys	[structure: H_2N CO_2H, $=O$, HO]	Tyr	[structure: H_2N CO_2H, indole ring]	Arginine	Asn	His
N	Arg	Y	His	Gln	Glu	Lys	Asp	Tryptophan
Asp	[structure: H_2N CO_2H, imidazole ring]	Trp	Lysine	Arg	Asn	Gln	Tyr	[structure: H_2N CO_2H, CO_2H]
Trp	Asn	[structure: H_2N CO_2H, O NH_2]	Arg	His	Asp	Tyr	Glu	[structure: H_2N CO_2H, $(CH_2)_4$ NH_2]
Tyr	D	Glu	Gln	Trp	Lys	Asn	H	Arg
[structure: H_2N CO_2H, NH, N NH_2]	Lys	His	Glu	Asn	Tyr	[structure: H_2N CO_2H, CO_2H]	Trp	Gln
[structure: H_2N CO_2H, O OH]	Trp	Asp	Tyr	[structure: H_2N CO_2H, NH_2]	Arg	His	Q	[structure: H_2N CO_2H, $=O$, H_2N]
K	Gln	Asn	W	Asp	His	E	Arg	Tyr
His	Tyr	R	N	Glu	[structure: H_2N CO_2H, O NH_2]	Trp	Lys	Asp

FIGURE 14.4 Amino acid Sudoku puzzle solved.

on a puzzle if there are going to be errors. Thus, to check the puzzle, one may frequently only spot-check several of the blanks (maybe a total of five or six is sufficient) to ensure they are entered correctly. A sudoku puzzle only has a single solution. The solution to the amino acid sudoku is merely the numerical solution given in Fig. 14.2 with each number substituted for the assigned amino acid. For convenience, the solution using three-letter abbreviations is shown in Fig. 14.4.

In the example shown, many of the decisions could have been made differently. One may choose different amino acids. One may choose to avoid one-letter abbreviations or structures. Such decisions can allow a more narrow focus for the student. Some of the burden necessary for creating the puzzle may be eased by removing certain features. The instructor may more easily construct the puzzle by choosing not to determine the easiest way to solve the puzzle, and thus not give this helpful information to students. The students will be faced with a more difficult task in such a case. Students who would be able to distinguish acidic from basic side chains would not be rewarded. However, the instructor would be able to devise the puzzles with less effort.

When this sudoku puzzle is assigned, some training of students is recommended. One possibility is, at the onset of the discussion of amino acids, to hand out the puzzle and to use it as one of the visual aids to point out the basics of amino acids. The instructor can show the correlation of the name, structure, and abbreviations and then follow up with a demonstration of the method and thought process used to solve the puzzle. For instance, as a class, one may complete all instances of lysine in the puzzle. Students who are unfamiliar with this type of puzzle would learn the strategy necessary. All students would better comprehend the necessity of identifying which structures, names, and abbreviations are correlated with each other.

EDUCATIONAL RESEARCH ON PUZZLES AND GAMES

Unfortunately, controlled studies concerning the effects of puzzles and games on learning are rare. More often, reports of educational game activities cite anecdotal evidence or informal surveys of student opinion on the value of these activities. This situation is unfortunate since well-designed studies could shed light on which kinds of activities may be best suited to a learning environment. Too often, however, experiment design is difficult as two populations of students (test subjects and control group subjects) are not conveniently available to measure the learning differences. As the number of chemistry games and puzzles has increased dramatically over the past 15 years, this indication of greater interest in these activities will hopefully lead to a more rigorous study of the effects of various puzzles and games. Effectiveness of these classroom activities is certainly an area that is ripe for pedagogical research.

CONCLUSIONS

Games and puzzles that support learning goals in the chemistry classroom are readily available for a wide variety of topics. The "fun" nature of teaching using a game may very well prove to be more effective than more traditional approaches. While still providing real learning, games and puzzles are frequently viewed as a classroom break, making students very receptive to this type of instruction. Whether one adapts existing material into a game format, applies a common game to chemistry material, or develops a novel game that applies specifically to chemistry, games have much to offer in the classroom. Many examples of games that can be made or purchased already exist, and only one's imagination need limit the development of further games that can be used to facilitate learning in the chemistry classroom.

REFERENCES

Bartholow, M. (2006) *J. Chem. Educ. 83*, 599.

Bertolini, T. M. and P. D. Tran (2006) *J. Chem. Educ. 83*, 590.

Capps, K. (2008) *J. Chem. Educ. 85*, 518.

Crute, T. D. A. (1992) *J. Chem. Educ. 69*, 559.

Crute, T. D. and S. A. Myers (2000) *J. Chem. Educ. 77*, 481.

Crute, T. D. and S. A. Myers (2007) *J. Chem. Educ. 84*, 612.

Deavor, J. P. (1996) *J. Chem. Educ. 73*, 430.

Edmonson, L. J. Jr. and D. L. Lewis (1999) *J. Chem. Educ. 76*, 502.

Erdik, E. (2005) *J. Chem. Educ. 82*, 1325.

Gottstein, G. (1998) Perplexors, Expert Level, MindWare, 1998 ISBN 1892069180

Grabowski, J. J. and M. L. Price (2003) *J. Chem. Educ. 80*, 967.

Greengold, S. L. (2005) *J. Chem. Educ. 82*, 547.

Hanson, R. M. and B. Michalek (2006) *J. Chem. Educ. 83*, 581.

Harsch, G. (1984) *J. Chem. Educ. 61*, 1039.

Helser, T. L. (2008) *J. Chem. Educ. 85*, 515.

Herron, J. D. (1996) *The Chemistry Classroom- Formulas for Successful Teaching*, pp. 18–19. Washington, DC: American Chemical Society.

Keck, M. V. (2000) *J. Chem. Educ. 77*, 483.

Koether, M. C. (2003) *J. Chem. Educ. 80*, 421.

Lemley, P. V. (1989) *J. Chem. Educ. 66*, 1011.

Most, C. Jr. (1993) *J. Chem. Educ. 70*, 1039.

Myers, S. (2003) *J. Chem. Educ. 80*, 423.

Nowosielski, D. A. (2007) *J. Chem. Educ. 84*, 239.

Olbris, D. J. and J. Herzfeld (1999) *J. Chem. Educ. 76*, 349.

Olbris, D. J. and J. Herzfeld (2002) *J. Chem. Educ. 79*, 1232 and references within.

Oppertshauser, K. (2008) http://classes.hdch.org/tkoppertshauser/Biology%203C/Microbiology/Jeopardy.2008.ppt (accessed December 18, 2008).

Perez, A. L. and G. Lamoureux (2007) *J. Chem. Educ. 84*, 614.

Peris, M. (2007) *J. Chem. Educ. 84*, 609.

Rosemond, J. (2008) http://www.rosemond.com/view/389/22010/9208.html (accessed October 17, 2008).

Russell, J. V. (1999a) *J. Chem. Educ. 76*, 487.

Russell, J. V. (1999b) *J. Chem. Educ. 76*, 481.

Saecker, M. E. (2007) *J. Chem. Educ. 84*, 577.

Samide, M. J. (2008) *J. Chem. Educ. 85*, 1512.

Scarpetti, D. (1991) *J. Chem. Educ. 68*, 1027.

Sevcik, R. S., O. Hicks, L. D. Schultz, and S. V. Alexander (2008a) *J. Chem. Educ. 85*, 514.

Sevcik, R. S., R. L. McGinty, L. D. Schultz, and S. V. Alexander (2008b) *J. Chem. Educ. 85*, 516.

Smith, D. K. (2006) *J. Chem. Educ. 83*, 1621.

Swain, D. (2006) *J. Chem. Educ. 83*, 589.

Welsh, M. J. (2007) *J. Chem. Educ. 84*, 610.

Yates, J. (2005) http://www.chessandpoker.com/sudoku-strategy-guide.htm (accessed December 4, 2009).

Zanger, M., A. R. Gennaro, and J. R. McKee Jr. (1993) *J. Chem. Educ. 70*, 985.

*Making Chemistry Relevant: Strategies for Including All Students in
A Learner-Sensitive Classroom Environment*, Edited by Sharmistha Basu-Dutt
Copyright © 2010 John Wiley & Sons, Inc.